ASPARTAME (NUTRASWEET★)
IS IT SAFE?

ASPARTAME (NUTRASWEET★) *IS IT SAFE?*

H. J. Roberts, M.D.

Œ

The Charles Press, Publishers
Philadelphia

Library of Congress Cataloging-in-Publication Data

Roberts, H. J. (Hyman Jacob), 1924–
    Aspartame (Nutrasweet★): is it safe? / by H. J. Roberts.
        p.   cm.
    Includes bibliographical references.
    ISBN 0-914783-37-8: $19.95
    1. Aspartame—Physiological effects.   2. Aspartame—Health
aspects.   I. Title.
QP801.A84R63   1990   89-81086
363.19′29—dc20   CIP

                        ISBN 0-914783-37-8
                        5 4 3 2 1

The Charles Press, Publishers
Post Office Box 15715
Philadelphia, Pennsylvania
19103

Designed by Sanford Robinson
Printed in the United States of America

CONTENTS

PREFACE

Simply stated, the theme of this book is that the well-known artificial sweetener, aspartame (known best by its trade name, Nutra-Sweet★), is potentially dangerous and may produce a wide variety of physical and mental symptoms, most of which now go unrecognized or are misinterpreted as serious illnesses.

The problem is far-reaching, I believe, involving probably thousands and thousands of unsuspecting victims. For this reason I've chosen to present the evidence and my views directly to the public by way of this book. In doing so, I have attempted to inform the reader about all pertinent aspects of the issue, including my personal observations as a physician who has examined scores of patients with reactions to aspartame-containing products. The challenge of condensing so many clinical and historical facts proved to be formidable, especially while trying diligently not to overwhelm the general reader with complex medical terms. I hope I have achieved a happy balance.

My comments and opinions have not been motivated by disrespect or malice toward the producers of aspartame and products containing it, or toward their representatives, and investigators. Trademark names have been omitted for general identification, and retained where they have been used in documents within the public domain (as in the *Congressional Record*). I have had no financial support from any company or institution having a vested interest within this realm.

ACKNOWLEDGEMENTS

I acknowledge with gratitude the assistance of the following individuals.

Elaine Airoldi, Shirley Brightwell, Beverly Ylen, Marjorie Childs, and Deborah Rose for their secretarial services.

Kellie McDaniel, Lillian Capote, George Leavitt and Jane Scotto of the Data Information Services at Good Samaritan Hospital (West Palm Beach) for computer assistance with the survey questionnaire.

Helen Musgrave, Sandra Pooler, Dorothy Gerrior and Betty Mangifesta for distributing the questionnaire.

Dr. David G. Hattan (Chief, Regulatory Affairs Staff, FDA Office of Nutrition and Food Sciences) and Dr. Linda Tollefson for furnishing computer printouts of aspartame-associated complaints submitted by consumers to their agency.

Pamela Roberts, Jackie Taylor, Linda O'Callaghan, Barbara Burke, and Claire Searl for assistance with the references.

Carla Santiago for drawing the formula of aspartame.

Esther Sokol and Stephen Roberts for their perceptive commentaries.

The hundreds of persons with apparent reactions to aspartame-containing products, and their relatives or friends, who completed the nine-page questionnaire deserve thanks for this considerable effort.

The following individuals, companies, institutions, journals and publishers kindly permitted me to reproduce pertinent excerpts, data,

and other figures or tables. The original sources appear in the text
and bibliography.

Annual Reviews, Inc.
Jeffrey L. Bada, Ph.D.
British Medical Journal
Journal of the American Medical Association
Journal of Pesticide Reform
The Lancet
Mr. Rodney E. Leonard
Marcel Dekker, Inc. (publisher of *Aspartame: Physiology and Bio-
    chemistry,* edited by L. D. Stegink and L. J. Filer, Jr., 1984)
Medical Tribune
Modern Medicine
Mrs. Barbara Mullarkey
New England Journal of Medicine
News America Syndicate (distributor of the *Arnold* cartoon by
    Kevin McCormick)
Mr. J. Stauber (artist for *Nebelspalter,* Zurich)
Charles C. Thomas (publisher of *Diseases of Medical Progress: A
    Study of Iatrogenic Disease* by R. H. Moser, M.D., 3rd edition,
    1969)
John B. Thomison, M.D. (editor, *Southern Medical Journal*)
    World Press Review
Richard J. Wurtman, M.D.

The support of earlier researches by Mrs. Muriel Brenner, Mr.
and Mrs. Louis Brizel, Mr. Robert Sanders, Mr. Merrill Bank, Mrs.
Bette Schapiro, Mr. Samuel Hausman, Mr. Edward Jaye, and Colo-
nel Frank Randolph is gratefully acknowledged, especially in the
absence of any governmental or corporate grant.

ASPARTAME
(NUTRASWEET★)
IS IT SAFE?

CHAPTER ONE

The Aspartame Boom

The beginning of wisdom is to call things by the right names.

— *CHINESE PROVERB*

ASPARTAME, KNOWN BEST BY the trade name Nutra-Sweet© *, is now consumed by more than 100 million persons in the United States. In addition to being 180–200 times as sweet as table sugar (sucrose), the virtual absence of calories in the chemical, coupled with the lack of cholesterol and salt, has enhanced its popularity.

Scores of products (over 1,200) contain aspartame. They range from soft drinks and tabletop sweeteners to puddings, gelatins, cereals, hot chocolate, gum, and palatable over-the-counter medicines for treating fever, headache and infection in children. The Food and Drug Administration (FDA) recently allowed its incorporation into fruit drinks, tea beverages, breath mints, frozen novelties, and flavored sprays to aid dieters. Furthermore, it is likely to be approved in the near future for baked goods, yogurt, wine coolers and low-alcohol beer.

The consumption of aspartame has increased at an astonishing rate since 1981, when this man-made sweetener was first introduced. By 1985, 800 million pounds of aspartame were used per year in the United States. The unparalleled acceptance is quite understandable in

*A registered trademark of the NutraSweet© Co.

a health-conscious society preoccupied with slimness and obsessive weight control.

Equally important, the public, increasingly concerned in recent years about all food additives, has been reassured that aspartame is entirely safe and can be used without worry. Indeed, aspartame is described as the "most thoroughly tested additive in history" . . . an allegation repeatedly made by manufacturer, the FDA, and other reputable persons or institutions. Consider just a few of the endorsements for the safety of products containing aspartame:

- The FDA Commissioner who approved aspartame on July 24, 1981 asserted that "few compounds have withstood such detailed testing and repeated close scrutiny." (46 *Federal Register* p.38289, July 24, 1981).
- Dr. Stanford Miller, Chief of the FDA's Bureau of Foods, told a Senate hearing: "I don't know of any substance in recent years that has been looked at with the intensity of aspartame. No one has yet come up with the slightest evidence to show that we were wrong in approving it" (*Congressional Record-Senate* May 7, 1985, p.S5493).
- The American Dietetic Association and the American Diabetes Association actively recommend the use of aspartame by diabetics as a "free exchange" (meaning that it need not even be counted as calories in the diet).
- Dr. Frank Young, current FDA Commissioner, told a Senate hearing on November 3, 1987: "In conclusion, we do not have any medical or scientific evidence that undermines our confidence in the safety of aspartame. This confidence is based on years of study, analysis of adverse reactions, and research in the scientific community, including studies supported by the FDA."
- Dr. Jean Mayer (President of Tufts University) and Jeanne Goldberg (1988) stated that aspartame has "a clean bill of health." They further asserted there is "no evidence that aspartame used in moderation poses any threat to health."

All of this sounds very convincing, doesn't it? Why, then, would I, a busy and established internist, take on the very formidable task of presenting an opposing view . . . arguing that aspartame is by no means as safe as we have been led to believe?

In fact, I believe that products containing aspartame are capable of producing, and reproducing, a wide spectrum of frightening symptoms, including severe headaches, convulsions, memory loss, and diarrhea—to name just a few.

My reasons for bringing this potential major health issue to the public's attention are explained throughout the book. Suffice it to say at this point that I am especially concerned about the desperate patients I and other physicians see who suffer from what seem to be obscure or elusive illnesses—when in fact the problem is related to usage of aspartame products.

Two questions may concern the reader right at the outset. First, "What information does this doctor have to justify stirring up anxiety about aspartame when the American Medical Association and dozens of regulatory agencies here and abroad have vouched that the additive is safe?" My answer: *"Read the book, weigh the evidence, and judge for yourself."*

The second question pertains to my credentials. Am I a reputable physician with first-class training and top-level credentials whose observations and opinions can be trusted? Briefly, I have been certified—and recertified—by the American Board of Internal Medicine, and am a fellow or member of several prestigious medical and scientific organizations. I have published more than 200 articles and letters in leading medical journals and have written five books. My first text, *Difficult Diagnosis: A Guide to the Interpretation of Severe Illness* (W. B. Saunders Co., Philadelphia, 1958), was used in its English, Spanish, and Italian editions by tens of thousands of physicians, both in their practice and to prepare for Board examinations. I practice internal medicine in West Palm Beach, Florida. Many of the patients I see in consultation have been referred for unresolved diagnostic problems, generally after seeing several other physicians and being studied at various medical centers and clinics. For what it's worth, I have been listed in *The Best Doctors in the U.S.* by John Pekkenen (Seaview Books, New York, 1981)—one of four Florida physicians listed under general internal medicine (and selected without my knowledge by other physicians.)

MY INTRODUCTION TO ASPARTAME

How did I get involved with aspartame in the first place?

I began to experience vague, but mounting, diagnostic misgivings early in 1984 for several reasons. As this uneasiness grew, I kept asking, "Why is my treatment for the *very same* types of complaints less effective now than it was in previous years?" I considered the toll of societal stress as a primary factor, but promptly discarded this notion. Another idea kept recurring: "Has some *new* factor been introduced into the arena of medicine?"

Over the next couple of years of professional soul-searching, I never considered the role of aspartame as a possible culprit. In fact, I regarded aspartame as a useful product at first. Having been interested in so-called reactive hypoglycemia ("low blood sugar attacks") for more than a quarter century, I welcomed the availability of any safe sugar-free product that could satisfy the sweet tooth of such patients ... and thereby improve their quality of life.

Tammy

The matter crystallized in my mind late one afternoon when Tammy, a 16-year-old girl, developed a recurrent seizure right in my office. She had been referred one month earlier for convulsions that failed to respond to conventional treatment prescribed by two competent neurologists. When I documented severe reactive hypoglycemia in her case, the crucial diagnosis seemingly was uncovered. But Tammy continued to suffer seizures. I can assure you that witnessing another attack under these circumstances was a humbling— as well as frightening—experience.

I carefully re-examined Tammy, but could find nothing different. Moreover, her blood glucose (sugar) concentration at the time of the attack proved perfectly normal; therefore, this particular seizure had not been provoked by hypoglycemia. Furthermore, her mother stated that Tammy took the mid-afternoon snack I had recommended to avoid a hypoglycemic attack. Specifically, she ate an aspartame-sweetened chocolate pudding.

At that point, the mother perceptively volunteered, "You know, Doctor, Tammy's grandmother has severe reactions to aspartame." I was stunned by this revelation. She then added that Tammy had been drinking more diet beverages during the preceding two weeks in order to avoid sugar . . . just as *I* had prescribed.

I have long maintained that a doctor's greatest teachers are his own patients. Their observations—and criticisms—are grounded in a reality that transcends contemporary dogmas ("the party line") accepted by "the establishment." I therefore suggested to the mother, "Let's see how Tammy gets along if she avoids all products containing aspartame."

When I saw her three weeks later, Tammy was doing fine. She felt much better generally. But most important, there had been no further seizures.

Deciding to pursue the question, I candidly told Tammy and her mother, "This dietetic pudding angle might be just a coincidence. But since so many products now contain aspartame, I think we should try to determine if it *really* did precipitate that attack in my office. So I'd like to propose that we give Tammy a serving of the *same* pudding and observe her as we did during the previous glucose tolerance test. Of course, I'll be watching her closely." The patient and both parents consented.

One morning shortly thereafter, Tammy ate the dietetic pudding in my laboratory. About two and a half hours later, she began to show signs of a seizure. I again anguished as the generalized muscle contractions, facial grimacing, severe confusion and sweating developed. Fortunately, prompt feeding of some milk and crackers aborted a full-blown convulsion. (It is of interest that Tammy's blood glucose concentration fell more at the second hour after ingesting the aspartame than it did when she previously swallowed the glucose solution.)

Tammy had no further seizures for nearly a year after abstaining from aspartame-containing products. As further proof of "the pudding connection," she successfully resumed competitive long-distance track.

The Rest of the Story

This experience with Tammy opened a Pandora's box of medical revelations. In questioning patients with hypoglycemia and diabetes, I was astounded by their prodigious consumption of aspartame products.

Other surprises followed. As the number of seeming aspartame-related problems in my practice escalated, I began asking *all* patients about their use of such products. But I had to be especially careful to avoid becoming trapped in the quicksand of a favorite new theory.

I therefore adopted certain fundamental criteria before labeling someone an "aspartame reactor." They included this triad:

(1) The individual began consuming products containing aspartame—or had recently increased such consumption dramatically—before the onset or aggravation of symptoms.
(2) The patient's complaints improved *dramatically* within several days or weeks after avoiding aspartame.
(3) The *same* set of problems *promptly* recurred after the patient was rechallenged with aspartame—whether knowingly (as by self testing) or inadvertently with the intake of products not suspected of containing aspartame (as at a restaurant).

All the while, I sought out every available pertinent piece of literature on aspartame. This included its pharmacology and physiologic effects, the background of its licensing and regulation, and published case reports of adverse reactions to aspartame products.

Increasingly, three considerations alarmed me. First, not enough was known about this synthetic chemical. Second, aspartame had been approved *without any extensive pre-marketing trials on humans.* Third, it did not come under intense FDA scrutiny because of its classification as a food "additive" rather than a drug.

This clinical and scientific pursuit of reactions to aspartame products unexpectedly became both a professional challenge and personal mandate. In an attempt to maintain total objectivity and autonomy, I did not accept research money from any individual, corporation, or institution having vested interests in the sugar, sweetener or drug industries. With only a shoestring budget, supported by a few patients and myself, it was slow going. (To set the record straight, I drew no salary for this research.)

By July 1986, I had accumulated data on *100 persons* who were apparent reactors to aspartame-containing products!

CHAPTER TWO

How I Became Involved in the Aspartame Issue

Two roads diverged in a wood, and I—
I took the one less traveled by,
And that has made all the difference.
— *ROBERT FROST*
(The Road Not Taken)

I BEGAN TO DELVE seriously into the subject of aspartame reactions. As I realized that they were accounting for some of the unexplained symptoms I had been encountering in patients, two bothersome thoughts kept surfacing. (1) *"Why was I a majority of one?"* (2) *"Why couldn't I find similar information about aspartame reactions in the medical literature?"*

It seemed that every source I turned to for valid medical facts either proved a dead end or simply reiterated the manufacturer's standard response—namely, these reactions were only *anecdotal* accounts of *rare* events.

I'll have more to say in later chapters about reasons for the scarcity of this information in medical books and journals. But let me consider briefly at this point the issue of "anecdotal" accounts. (Those who dispute my views about the potential dangers of aspartame inevitably invoke the argument that much of my case is based on material obtained from my patients and correspondents, notwithstanding its detailed nature.)

The medical profession usually pays little attention to anecdotal reports. After all, an anecdote is only a story, especially if not backed by "real" state-of-the-art scientific evidence. Say, for example, that five different patients tell me they develop headaches each time they drink a soda beverage containing aspartame. Most scientists would ignore or minimize the possible relationship on this basis alone because the patient's accounts are only anecdotal.

To prove that aspartame was really at fault, scientists would insist on a *double-blind study*. Under these conditions each patient would drink a soda, chosen at random, which may or may not contain aspartame. Neither the patient nor the investigator know which are the aspartame-containing sodas (hence the term double-blind). Should headaches develop following consumption only of the aspartame beverages, the evidence is considered secure.

Does this mean that all anecdotal evidence is worthless? Not at all. Consider the following comment by Dr. Charles Harris (1987):

> "But the medical profession has a tendency to discard out of hand, and disparagingly, "anecdotal" information. Digitalis, morphine, atropine, and the like are chemical derivatives that stem from anecdotal folklore remedies. After all, one anecdote may be a fable, but 1,000 anecdotes can be a biography ... A vital function of the medical profession is to sift the anecdotes and submit them, if possible, to scientific evaluation. But it all starts as anecdote."*

Recognizing the inherent problem of proving my case, I moved cautiously. In fact, I probably would have dropped the matter right there had my background been more conventional.

But it wasn't.

I felt that aspartame might represent a serious health threat, and had to find out. Several times, however, I nearly convinced myself about having gone too far out on a limb. But invariably, something would direct me back to aspartame reactions.

A few of the forces that influenced my thinking may be of interest to you.

---

*© 1987 *Medical Tribune.* Reproduced with permission.

INFORMATION FROM PATIENTS

Again and again, I heard negative clinical comments about aspartame products. I began to *routinely* question patients about them, and was surprised by the frequency of such complaints. An example: "Oh, I've long since stopped using the stuff because it makes me terribly sick! I didn't mention it to you before because I thought you'd consider me a hypochondriac."

I came to admire the astuteness of patients who had deduced, on the basis of fairly respectable self-research, that aspartame products caused or contributed to their problems. Some scrutinized diaries of everything they did, ate, and drank. A few determined parents of children suffering from unexplained headaches, seizures, learning disabilities, or hyperactivity demonstrated lots of creative scientific imagination . . . including in-house elimination diets. One mother stated:

> "Since we were going to try the 'nutrition route', we decided on our own to take him off the aspartame drink he had been consuming for the past four months (ever since I had been on a diet) . . . He was drinking four to six glasses a day . . . (Usually one glass was in the form of a frozen popsicle) . . . He has been a healthy little boy since, with absolutely no recurrence of any of the previous problems . . . It's been five months since the nightmare of the last seizure."

Some patients become aware that their symptoms disappeared when, for one reason or another, they stopped using aspartame.

> The fiance of a young woman had agonized with her during a one-year ordeal of unexplained extreme headaches. Numerous consultations and multiple therapies (including Percodan® for pain) failed to provide relief. On one occasion, they happened to transpose their beverages—her glass containing a diet cola, and his a regular cola. It was the first time he had tasted an aspartame beverage. Feeling "lousy" shortly thereafter, he urged her to avoid diet colas. The headaches diminished within a few days . . . and then disappeared.

There were comparable "Eureka" experiences when *complaints van-*

ished shortly after abstinence from aspartame due to a change in routine. The most frequent scenario involved feeling well while traveling abroad, where aspartame products were not easily purchased—only to suffer a prompt recurrence on returning home and resuming their use.

REPORTS FROM FORMER FDA PROFESSIONALS

I encountered several reports from former FDA professionals expressing concern over the safety of aspartame.

- Dr. Martha Freeman (previously with the FDA Bureau of Drugs) concluded a memo in 1973: "The information provided is inadequate to permit an evaluation of potential toxicity of aspartame."
- An FDA Special Task Force rendered a similar opinion.
- Dr. M. Adrian Gross, a respected FDA scientist-pathologist, listed *dozens* of "serious deficiencies" by a manufacturer and its representatives concerning the prelicensing evaluation of aspartame in animal studies . . . the most serious pertaining to brain tumors.

SENATOR HOWARD METZENBAUM

I learned that two U.S. Senate hearings on aspartame had been held at the prodding of Senator Howard Metzenbaum of Ohio—specifically, on May 7, 1985 and August 1, 1985. As I read the official proceedings, I kept rubbing my eyes in near disbelief. (As an aside, strong glasses or a magnifying glass are recommended for persons not used to reading the *Congressional Record* fine print.) Noting that over 20 billion cans of aspartame soft drinks would be consumed during 1985, Senator Metzenbaum observed:

"We had better be sure that the questions which have been raised about the safety of this product are answered. I must say at the outset, this product was approved by the FDA in circumstances which can only be described as troubling."

The Senator subsequently reiterated that he was not aware of any extensive studies *on humans* prior to FDA approval of aspartame and its marketing! I'll have much more to say about the Senate hearings later.

REPORTS FROM OTHER SOURCES

I chanced upon isolated reports in the medical literature that incriminated aspartame-containing products as a cause of convulsions (Chapter 9) and hives (Chapter 18).

Respected journalists decried the "flawed tests" on the basis of which aspartame had been approved for commercial use. Florence Graves (1984), Editor of *Common Cause,* wrote a scathing critique on this subject.

At least, I wasn't alone in my views.

ASPARTAME VICTIMS AND THEIR FRIENDS

While going through some research files, I came across a newspaper reference to Aspartame Victims and Their Friends. It was a consumer group founded by Shannon Roth of Ocala, Florida. This young woman steadfastly attributed her permanent blindness in one eye to the heavy consumption of aspartame-containing products.

Out of curiosity, I called Shannon to confirm the newspaper account. I nearly fell out of my chair when she informed me that *more than 800* "aspartame victims" had called on her "hot line." The longer we talked, the more convinced I became that (1) neither Shannon nor most of her correspondents were neurotic or paranoid; (2) the magnitude and severity of aspartame reactions were not coincidental "anecdotes" or rare "idiosyncracies"; and (3) a possible imminent public health threat might be in the offing.

CONSUMER COMPLAINTS

I tallied the number of formal complaints *volunteered* by consumers with alleged reactions to aspartame products. These had been sent to the FDA, the Centers for Disease Control (CDC), the manufacturer, Aspartame Victims and Their Friends, other consumer organizations, and interested investigators. To my amazement, the figure as of mid-1986 exceeded *10,000* individuals—with the admitted reservation that some of the same complainants could have been included in the data of more than one group.

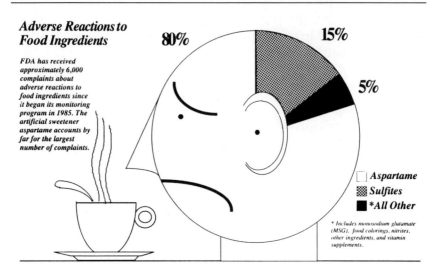

Adverse Reactions to Food Ingredients

FDA has received approximately 6,000 complaints about adverse reactions to food ingredients since it began its monitoring program in 1985. The artificial sweetener aspartame accounts by far for the largest number of complaints.

80%   15%

5%

☐ Aspartame
▨ Sulfites
■ *All Other

* Includes monosodium glutamate (MSG), food colorings, nitrites, other ingredients, and vitamin supplements.

FIGURE 2-1
FDA Consumer, October 1988, p. 17

Complaints received by the FDA were of particular interest. In 1985, the agency started a monitoring program to collect information about adverse reactions to food ingredients. **Of 6,000 complaints received, 80% concerned aspartame.** Figure 2-1 shows the findings in a dramatic manner.

Were 4,800 individuals (80% of 6,000 complaints) entirely wrong? Was this "anecdotal" information worthless? Not in my view. Something was amiss about aspartame and products containing it.

MEDICAL COSTS

I was distressed by complaints about the extraordinary costs that patients had incurred in pursuing the cause of their symptoms. The large number of physicians and consultants who evaluated these "problem patients" *before* the possibility of an aspartame reactor was raised proved astonishing. Dozens of such persons had seen the *combination* of neurologists, ophthalmologists and allergists . . . but to no avail. The associated costs for the prior consultations, testing, and hospitalization(s) were staggering. Consider the case of one of my patients.

I saw a 29-year-old businessman in consultation for repeated grand mal convulsions. The first seizure occurred four months after he began drinking aspartame beverages. Prior to my visit, he already had *five* CT scans of the head, *three* magnetic resonance imaging (MRI) studies of the brain, *five* electroencephalograms (EEGs), and several hospitalizations!

THE TOLL ON PATIENTS

The drastic effects of aspartame-associated reactions on the lives of patients and their families disturbed me greatly. The toll included personal, occupational and financial losses. The extent of these problems can be judged from a survey questionnaire I later used in studying aspartame reactions (see Chapter 8). The estimated impact on their lives was classified as "a bit"—47%; "severely" (such as loss of a job)—17%; "I was incapacitated":—10%.

A 35-year-old male anesthetist had three grand mal seizures, severe headaches and visual difficulty while drinking 4-6 diet colas daily, but none for two years after stopping aspartame. He told the U. S. Senate hearing on *"NutraSweet"—Health and Safety Concerns,* held on November 3, 1987:

"My problems are now reobtaining my malpractice insurance and getting my privileges back. I have only one small hospital where I could practice and one plastic surgeon's office. As of yet, I have not been reinstated with malpractice insurance" (p.304).

A number of individuals had their driver or pilot licenses revoked or suspended because of aspartame-associated impaired vision, convulsions, or confusion—depriving them of employment.

A young Air Force pilot told the Senate hearing held on November 3, 1987 that he suffered a grand mal seizure while consuming up to one gallon of an aspartame beverage daily. There had been no recurrence over the ensuing two years of abstinence. Nevertheless, he was permanently grounded because of the diagnosis of an "idiopathic partial seizure disorder" (Collings 1988).

CONTRIBUTORY FACTORS

Other facts intensified my professional concern, particularly the phenomenal consumption of aspartame-containing products by the public.

- The diet beverage market, clearly the fastest growing part of the soft-drink industry, was accounting for 24% of soft drink sales, and increasing at a rate of 20–25% annually. At this rate, it is estimated that diet soft drinks will outsell sugar drinks early in the next century (*The Wall Street Journal* January 24, 1989, p.B-1).
- An aggressive marketing campaign by one producer pitted a diet cola against a *non*diet leader for the first time (*The Wall Street Journal* January 24, 1989, p.B-2).
- Consumers began drinking up to six times as many diet drinks as those using sugared sodas (*The Wall Street Journal* June 27, 1988, p.20).
- Consumption seems destined to spiral following FDA approval of the incorporation of aspartame in fruit drinks, tea beverages, baking mixes and a host of other products.
- The lowered costs of aspartame after the expiration of patent rights in the United States (during 1992) probably would herald more widespread use.
- The consumption of low-calorie drinks in Europe recently surged, especially when France lifted its ban on such beverages (Kamm 1988). Spending for one popular aspartame-containing beverage more than tripled in Britain during 1988 (Kamm 1988). The so-called "cola wars" undoubtedly will intensify with finalization of the European Common Market, thereby allowing for more efficient international distribution of these products under a single management structure.

Anxiety was fueled by the aggressive promotion of aspartame-containing products in the media. It probably constitutes the largest advertising campaign ever designed around a product ingredient—more than $100 million a year. This was evidenced by the vast sums spent by one aspartame cola producer for repeated television ads during the 1989 Super Bowl football game . . . including the three-dimensional half-time show as a "first." Advertisers for this event paid *$675,000* for a *half-minute* spot (*The Wall Street Journal* January 20, 1989, p.B-4)!

- Celebrities paid to make this pitch have included Bill Cosby, Raquel Welch, Billy Crystal, Joe Montana, Dan Marino, Marvin Hagler, former Democratic Vice-Presidential candidate Geraldine Ferraro, and boxer Mike Tyson (the "undisputed champion" in the diet cola competition.) New media stars are quickly engaged to plug diet drinks and products— even Roger Rabbit (the animated character from the movie *Who Framed Roger Rabbit?*)
- The "battle of the blimps" at major sports events now features the names of popular aspartame-containing soft drinks on such aircraft.
- A sophisticated media campaign gives the impression that aspartame is a "natural" substance (rather than man-made) substance. Dr. Thomas E. Lindow (1988) aptly averred that most of the health-care "information" given by industry to the public is basically advertisement material aimed at enticing rather than informing.

A 30-year-old female who worked for two radio stations had been experiencing severe confusion, impairment of memory, depression with suicidal thoughts, insomnia and dizziness. She used one packet of an aspartame tabletop sweetener to each cup of coffee. These complaints dramatically improved within three days after stopping the aspartame—and did not recur. She expressed her indignation in these terms: "I become angry when I see commercials promoting aspartame. How harmless they make it all look. The All-American brown-and-white can and something about bananas. If this product is truly harmless, why the marketing aspect?"

- Many TV commercials featuring household-name stars are targeted to child viewers. A major manufacturer of aspartame-containing soft drinks has already taken aim at children watching Saturday morning television, known as the "moppet market" (*The Wall Street Journal* December 9, 1988, p.B-1). Producers had previously avoided such targeting due to concern over the unique vulnerability of children, and the possible exacerbation of obesity among sedentary young video viewers. This pitch at second generation "baby boomers" may herald a shift to diet drinks when a certain age and waist size are attained.

- A public relations blitz is aimed at our weight-conscious public, especially young women with the "fear of fat." The six-page "Guide to Health & Fitness" insert for a popular table-top sweetener emphasized its value in "slimming down" and the change of "your waistline!"

- Another campaign encourages young adults to replace breakfast coffee with diet colas. Wiedner and Istvan (1985) reported that soft drinks had become the primary source of dietary caffeine among persons in the 18-to 24-year-old bracket.

- Multicolor ads in medical, nursing and dietetic journals feature these appealing low-calorie, salt-free, caffeine-free and cholesterol-free products.

THE "BOTTOM LINE"

I could not ignore the most pressing consideration of all; the consumption of a potentially harmful chemical additive by more than 100 million persons in the United States. Even if only *one-tenth of one per cent (0.1%)* suffered adverse reactions to aspartame-containing products, a highly conservative estimate, my calculation came to 100,000!

A CONCERNED PHYSICIAN STEPS FORWARD

As my concerns about aspartame grew and I became convinced that this additive adversely affected a substantial number of individuals, I had to decide whether to follow customary medical protocol in expressing my views or take a bolder step and bring this matter directly to the attention of the public.

The usual approach involves presenting a "paper" at a medical meeting or, more often, submitting a manuscript to a journal for publication. In the case of aspartame reactions, both of these alternatives were likely to face many obstacles . . . with months and months elapsing before they could ever be resolved. As discussed previously, my data about aspartame would immediately be categorized as anecdotal and, therefore, rejected outright or given very low priority for acceptance by peer groups and editors.

This is not to say I ignored normal medical channels in presenting my findings. In fact, I submitted my evidence to various journals—largely without success. At the time of writing of this book, however, a major article was published in a peer-reviewed periodical (Roberts, H. J.: Reactions Attributed to Aspartame-Containing Products:551 Cases. *Journal of Applied Nutrition* Vol. 40, 1988, pp.85–94). I also presented my views at important annual medical and scientific meetings (for example, the Southern Medical Association and the American Association for the Advancement of Science).

A physician who takes pride in his long commitment to medical science and clinical research doesn't go public (as they say) with new observations just willy-nilly. The possibility that colleagues might misconstrue good intentions as self-serving sensationalism was a source of worry and anguish for me.

Despite these concerns I felt convinced that the public deserved to be informed of contrary opinions about aspartame and products containing it. For this reason I decided finally to express my views by way of the media and this book.

My media appearances, although hardly frequent, generated remarkable interest as judged by the deluge of letters and telephone calls I received. The responses were polarized. As anticipated, certain organizations and individuals having vested interests criticized me severely on the basis of (1) the "anecdotal" nature of my case material, (2) the absence of double-blind studies, and (3) my "chemophobic" attitude (meaning, I suppose, that I have a reflexive aversion to all chemical additives). One critic went so far as to charge me with "media terrorism."

On the other hand, I was delighted with the many grateful letters and calls from aspartame reactors and their families. One Louisiana correspondent wrote:

> "I want to thank you for speaking out about aspartame causing medical problems . . . When a mere patient tries to speak out to defend his health against chemical additives, he meets deaf ears and closed doors. But when a doctor such as yourself is brave enough to speak out for us, we are encouraged that there may yet be a chance to get the chemicals out of our food. Thanks again. Please don't let 'them' silence you."

This reference to "them" caused me to sit up and *really* take notice. How did other physicians respond to my public appearances? Although I have not received any major acknowledgement thus far from a medical organization, many individual doctors have written to me about suspected aspartame reactions in their patients or members of their own family.

- A Philadelphia physician wrote me after reading about my first press conference: "I have had a few patients myself with seemingly the same complaints that have improved with removal of aspartame."
- A London physician described the otherwise-unexplained severe acidosis in his 9-year-old daughter after she drank an aspartame-containing cola (see Chapter 6).

I was no longer "a majority of one."

CHAPTER THREE

Aspartame Consumption

*The thing that bugs me is that the people think
the FDA is protecting them—it isn't. What the
FDA is doing and what the public thinks it's
doing are as different as night and day.*
— *D R. H E R B E R T L. L E Y (1969)*
Former FDA Commissioner

THE FOOD AND DRUG Administration (FDA) approved
aspartame for use in dry form on July 15, 1981, and as a low-calorie
(non-nutritive) additive to carbonated beverages on July 8, 1983.

In-house FDA scientists, as well as concerned investigators, physicians and consumer groups, protested such approval. Their reservations and testimony appear in the *Congressional Record-Senate*
proceedings of May 7, 1985 and August 1, 1985, and in a report by
the United States General Accounting Office (*Food and Drug Administration: Food Additive Approval Process Followed for Aspartame,* GAO Report B-223552, June 18, 1987).

A Public Board of Inquiry (PBOI) advised against the approval
of aspartame on September 30, 1980 pending further testing relative to induced brain tumors. It asserted: "The Board has not
been presented with proof of a reasonable certainty that aspartame (NutraSweet) is safe for use as a food additive under its
intended conditions of use."

The FDA denied requests for a further hearing on the incorporation of aspartame into carbonated beverages and carbonated beverage syrup bases (*Federal Register* 48: July 8, 1983, p.31382). It then gave final approval for such use (*Federal Register* February 22, 1984, pp.6672–6682).

ASPARTAME CONSUMPTION

Aspartame products are currently being consumed by more than 100 million persons in the United States . . . often in large amounts.

- During 1985, 400,000 tons of aspartame were used—averaging 5.8 pounds per person. This was the equivalent of 1.6 *billion* pounds of sugar.
- According to the Department of Agriculture, the per-capita consumption of artificial sweeteners (including aspartame and saccharin) rose from 6.2 pounds in 1975 to 17 pounds in 1985.
- Aspartame products held 70 percent of the artificial sweetener market as of 1989 (*The Wall Street Journal* February 6, 1989, p.B-1).

This "marketing miracle of the 1980s" represents the coupling of successful PR campaigns for aspartame with negative consumer attitudes toward sugar, saccharin, and cyclamate.

- Many aspartame reactors indicated that they were introduced to such products after receiving complimentary purchase coupons, especially for "diet" drinks.
- Aspartame products have deeply penetrated eating establishments and fast-food chains.
- The profits of several major gum producers were revitalized once they incorporated aspartame in chewing gum.
- Increased consumption of the two leading brands of aspartame-containing colas during 1986 allowed them to be ranked #3 and #4 among beverages consumed in the United States (*The Wall Street Journal* June 3, 1987, p.33).
- The 1986 Annual Report of the Coca-Cola Company noted that the sales of its diet cola increased 14 percent . . . thereby making it the third largest selling soft drink in the country.

- Sales of the major brands of diet cola grew five times faster during 1988 than those of regular colas (*The Palm Beach Post* July 18, 1988, p.15).
- Industry surveys allegedly revealed that 2 million American households stopped buying one major brand of a sugared cola beverage during 1988 (*The Palm Beach Post* January 20, 1989, p.14).
- Aspartame use has increased worldwide. For example, Canada was consuming an estimated 300 metric tons even before it went off patent.

Prodigious Consumption

Some individuals consume extraordinary quantities of aspartame products. The intense thirst induced by such products (see Chapter 21) may contribute to the prodigious amounts consumed by some people.

Some of my patients did not realize the magnitude of their intake of aspartame-containing products until I asked them—or their spouses—about the *actual* amounts being taken.

- A 29-year-old man with seizures drank *eighteen* 12-ounce cans of an aspartame cola beverage daily for one year.
- A 35-year-old woman complained of "gas," constipation (attributed to a "spastic colon"), headaches, problems with memory, and increasing anxiety. She was ingesting 10–15 cups of diet soda, up to 15 glasses of presweetened tea, and considerable amounts of diet cranberry juice daily.
- A 35-year-old pilot with convulsions, severe drowsiness, slurred speech, severe depression, irritability and intense headaches evidenced dramatic improvement within two weeks after stopping aspartame products. Until then, his *daily* consumption included:

  - Two to four 16-ounce bottles of aspartame soft drinks
  - One 2-liter bottle of aspartame soft drinks
  - Up to 12 packets of an aspartame tabletop sweetener in hot tea
  - One or two regular glasses of an aspartame beverage mix
  - Two to four glasses of an aspartame hot chocolate

- Up to two servings of aspartame-sweetened puddings and gelatins
- Several sticks of aspartame chewing gum

- A 34-year-old receptionist became incapacitated as a result of 21 major complaints, which promptly subsided after stopping aspartame-containing products. They included marked visual changes, ringing in the ear, severe headaches, intense dizziness, two convulsions, gross impairment of recent memory, severe depression, anxiety attacks, abdominal pain with severe nausea, vomiting, bloody diarrhea, and considerable thinning of the hair. Her *daily* intake had included:

  - Six 12-ounce cans of aspartame cola beverages
  - One 2-liter bottle of aspartame soft drinks
  - Five to six packets of an aspartame tabletop sweetener
  - Several glasses of presweetened iced tea
  - Eight glasses of aspartame hot chocolate
  - Eight servings of aspartame puddings and gelatins
  - 25-40 sticks of aspartame gum
  - 12-24 packs or teaspoons of other aspartame-containing products

Simple calculation makes it evident that these patients—and others like them—ingest large amounts of the amino acids phenylalanine and aspartic acid, and of methyl alcohol, which result from the breakdown of aspartame in the body (*see* Chapter 5 and 6).

Economic Impact

The economic impact of aspartame has been enormous. Its sales (in excess of $750 million during 1986) promise to spiral as new uses are permitted (see below). By the 1990s, the intake of aspartame is predicted to approach one-fourth the overall consumption of sugar (Pardridge 1986)!

Decreasing consumption of refined sugar provides a related perspective. The following per-capita figures from 1977 to 1987 were compiled by the Economic Research Service and the U.S. Department of Agriculture.

Year	Refined Sugar (in pounds)
1977	94.2
1978	91.4
1978	89.3
1980	83.6
1981	79.4
1982	73.7
1983	71.1
1984	67.6
1985	63.3
1986	60.8
1987	62.1

On this basis, it is evident that sugar consumption fell by one-third in just a decade.

A commercial attempt to counter this trend has been the introduction of a blend of cane sugar (99%) "sweetened" with aspartame (1%).

Extended Uses

More young adults are now drinking aspartame-sweetened colas rather than coffee in the morning. Sophisticated ads aimed at "the Pepsi generation" reinforce such preference for "a cool refreshing drink to come alive."

The FDA has since approved the use of aspartame in juice drinks, frozen novelties, tea beverages, breath mints, and other products—including lucrative weight-control "supplements" sold by franchised outlets and doctors specializing in obesity (bariatricians). The prior anticipated total annual commercial sales of $6 billion (*The Wall Street Journal* November 26, 1986, p.9) are probably conservative.

In addition, recent reports indicate that certain popular aspartame-containing "fiber" (hydrophilic mucilloid) preparations (for controlling constipation) also can lower the serum cholesterol in persons with elevated cholesterol levels (Anderson 1988). Cholestyramine, an important drug prescribed for this reason, also is available with aspartame to increase its palatability. Given the current extent of excessive fear over cholesterol (cholesterolophobia), the consumption of these products is increasing.

Finally, the announcement by the NutraSweet Company of its "new all-natural fat substitute" could herald another dramatic extension of aspartame consumption if both were to be combined and marketed as simulated ice cream. Projecting the potential for retail sales up to $15 billion, the Chairman asserted, "We've taught a protein to fool the mouth" (*The Wall Street Journal* January 28, 1988).

> On February 17, 1989, I submitted recommendations to the FDA (Roberts 1989a) as requested public comment (GRASP 8GO345). I urged extreme caution before licensing the "microparticulated egg and milk protein product" in the absence of extensive pre-marketing studies and trials in *humans* . . . lest the aspartame experience be repeated.

AVERAGE DAILY CONSUMPTION AMONG INDIVIDUALS WITH SYMPTOMS

Results from Questionnaire

As part of my investigation of aspartame, I sent a survey questionnaire to 1177 persons with aspartame-associated complaints to determine the average daily consumption of aspartame products when their symptoms began. The questionnaire and the full findings of the study are described in Chapter 7.

There was a 30.8% (!) response from this single mailing. A computerized analysis of aspartame consumption shows:

- 61 percent consumed two 12-ounce cans of an aspartame soft drink.
- 43 percent consumed three packets of an aspartame tabletop sweetener.
- 26 percent consumed two glasses of aspartame-sweetened iced tea.
- 13 percent consumed one glass of aspartame hot chocolate.

- 15 percent consumed one serving of aspartame pudding or gelatin.
- 12 percent consumed four sticks of aspartame gum.
- 7 percent consumed one bowl of an aspartame-sweetened cereal.
- Many responders used two or more aspartame products each day.

One out of two reactors indicated that they had *greatly* increased their intake of aspartame-containing products prior to the onset of complaints.

"Aspartame Is Everywhere"

The enormous consumption of aspartame-containing products correlates with their universal presence in food markets, other stores, and vending machines. Aspartame has become ubiquitous.

A 48-year-old woman with aspartame-related complaints wrote:

"It is becoming increasingly difficult to avoid aspartame. I am constantly irritated by people *not informing me* when a product containing aspartame is served. This has happened time and again at church suppers and friends' homes. *I do not want it foisted on me.*"

Many reactors suffered aspartame-associated attacks after unknowingly eating or drinking aspartame products in restaurants.

Another frequent oversight pertains to "health products" containing aspartame. They include vitamins, over-the-counter pain relievers, laxatives, and a host of other substances. For example, several of my patients who presented with unexplained severe abdominal pain, nausea, vomiting, diarrhea, and even bloody stools initially denied taking aspartame. Further questioning, however, revealed the use of a popular "sugar free effervescent natural fiber laxative" that contained aspartame (see above). Their complaints disappeared shortly after shifting to the non-aspartame form of this product.

CHAPTER FOUR

Aspartame Metabolism

*Medical science often cannot give adequate answers
to questions of toxicological action, not only as
regards new compositions but even in the case of
substances long in use. Indeed, to the medical man
unpleasant surprises are constantly being revealed,
in the case of both drugs and of chemicals used in
food preparations.*

— *SIR EDWARD MELLANBY*
*(1951)**

I HAVE NO DOUBT that most readers of this book will be far
more interested in the various clinical manifestations of aspartame
reactions than in the mechanisms that may induce those adverse effects. Not to worry, the greater part does indeed focus on the symptoms of aspartame reactions.

Nevertheless, I believe it is worthwhile to present a few more
details about aspartame so that you will be able to assess the scientific
merit of my concern. Let me begin by briefly reviewing the metabolism of aspartame—in other words, what happens to it in the body
after being swallowed.

CHEMISTRY

Aspartame is a *synthetic* chemical. Stated differently, it is not a

*© 1951 *British Medical Journal.* Reproduced with permission.

"natural" or "organic" product derived from banana plants or cows . . . as might be inferred from printed advertisements and TV commercials. Rather, it is a man-made substance.

Two of its three components are amino acids (the "building blocks of protein")—*phenylalanine* and *aspartic acid. Methyl alcohol,* the third component, is also known as methanol and wood alcohol.

The chemical formula for aspartame appears in Figure 4-1.

ASPARTAME

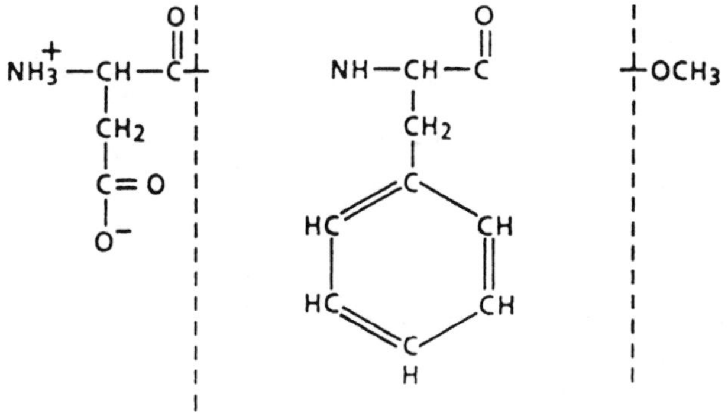

ASPARTIC ACID   PHENYLALANINE   METHYL ESTER (METHANOL)

FIGURE 4-1

Phenylalanine constitutes about 50 percent of the aspartame molecule, aspartic acid 40 percent, and methyl alcohol 10 percent. Expressed in simple quantitative terms, one liter (approximately a quart) of most aspartame-containing drinks contains about 550 mg aspartame. In turn, this yields 275 mg phenylalanine, 220 mg of aspartic acid, and 55 mg methanol. The amounts of aspartame can vary widely, however, in different beverages.

- Some brands of "diet" colas may contain as much as four times more aspartame than others.
- Orange-flavored beverages tend to have a higher content— up to 333 mg aspartame per 12 fluid ounces, or 930 mg per liter (*Federal Register* February 22, 1984, p.6677).
- Additional aspartame may be added to compensate for alteration of the taste of some sodas because of breakdown prod-

ucts that develop during storage, especially in a hot environment.

METABOLISM

After entering the body, the components of aspartame are promptly released within the upper gastrointestinal tract by enzymic action. They then are transported across the wall of the small bowel, and finally into the general circulation.

Methanol

This component is apparently the first to be separated from the aspartame molecule during digestion. As mentioned, methanol is a synonymous with *wood alcohol*, a deadly poison even when consumed in relatively modest amounts. Even in lesser quantities, methanol is potentially dangerous because the body is unable to detoxify it (unlike the *ethyl* alcohol found in whiskey, beer and wine, which is readily detoxified and excreted from the body).

The potential adverse effects of methanol—particularly on the brain, optic nerves, and retina—are described in later chapters.

Phenylalanine and Aspartic Acid

Promotional material for aspartame implies that the body treats aspartame's two amino acids no differently than if they were derived from fruit, vegetable, milk or meat. During his testimony before the U.S. Senate hearing on aspartame on November 3, 1987, Dr. Frank Young, FDA Commissioner, emphasized that the amino acids in aspartame are metabolized the same way as "natural" building blocks of protein.

I disagree. *There are profound differences in both the rates of digestion and the degree of absorption depending on whether these amino acids are provided by food or by aspartame.* The usual forms of protein (such as meat) contain four or five percent phenylalanine . . . *not* 50 percent, as does aspartame. Moreover, they are slowly digested within the gastrointestinal tract, and in tandem with other neutral amino acids. This precludes an unbalanced flooding of the system with single amino acids. The situation is quite different with aspartame, where the body is suddenly deluged with large amounts of two amino acids.

The administration of aspartame as capsules in double-blind studies or other provocative tests raises other technical considerations. For example, the concentrations of phenylalanine and aspartic acid (aspartate) in the blood are lower when three grams of aspartame are given in capsule form than when it is ingested as a solution (Stegink 1987). Furthermore, the peak blood levels of phenylalanine are more variable after capsule administration—ranging from 45 to 240 minutes—than after drinking aspartame in solution (Stegink 1987).

The serious consequences of excessive phenylalanine and aspartic acid suddenly flooding the blood stream and brain are explained in the next chapter. Special attention is directed to imbalances of the major amino acid-derived neurotransmitters that control nerve impulses, and influence both brain and endocrine function.

CHAPTER FIVE

Excessive Intake of Phenylalanine and Aspartic Acid

One really can't tell from animal testing how safe an additive will be in man. So the prudent thing to do, on the basis of the experimental data, is to lean over backward not to greatly increase risks.
— *DR. PAUL E. JOHNSON (1970)*

ASPARTAME, AS NOTED, CONTAINS the amino acids phenylalanine and aspartic acid. These are just two of the 22 amino acids the body normally uses for building proteins.

HOW THE AMINO ACIDS IN ASPARTAME CAN PRODUCE REACTIONS

After ingestion there is a rapid breakdown and absorption of aspartame within the upper gastrointestinal tract. This results in a prompt rise of the two amino acids in the bloodstream. The increase is greater than after the digestion of dietary proteins, which is relatively slow. Moreover, the so-called neutral amino acids in meat and other proteins tend to "neutralize" one another so that there is no sudden flooding of the bloodstream by one or two amino acids.

As people become older, they require lesser amounts of amino acids since the requirements tend to decline sharply as persons age

(Stegink 1984). Accordingly, eating or drinking large amounts of aspartame could pose a particular problem for older individuals (Chapter 28).

Two other factors probably contribute to aspartame reactions. First, the brain has a limited ability to adapt to a sudden unbalanced flooding with these amino acids, particularly phenylalanine.

Second, alterations of the physical structure of these two amino acids may occur during prolonged storage and exposure to heat. These changes can affect how the body handles aspartame biologically, and may lead to adverse effects. Specifically, the aspartame molecule and its amino acids can be transformed under these circumstances into structures that are mirror images, called *stereoisomers*, of the original substance. Particular attention is directed to the uncommon D (dextro) amino acids stereoisomers, which behave differently than the common L (levo) molecules in the body. A well-known clinical example of the difference in action of stereoisomers is the drug L-Dopa (levodopa) used as standard treatment for Parkinson's disease. In contrast, its counterpart, D-Dopa, has no effect.

The effects of entry of aspartame breakdown products into the brain may be even greater than currently believed. Studies on the neurotoxicity of aspartate and glutamate, a related compound, have demonstrated that several regions of the brain lack a so-called *blood-brain barrier*, which inhibits the passage of certain substances into the brain. As a result, aspartame and other substances, including monosodium glutamate (MSG), can penetrate the brain freely, and selectively exert toxic effects (see below) (*Environmental Health Perspectives* 26;107–116, 1978).

Dr. John W. Olney, Professor of Neuropathology at Washington University School of Medicine, wrote Senator Howard Metzenbaum on December 8, 1987:

"If glutamate and aspartate are released from cells and not rapidly taken back up, they flood the excitatory receptors on the external surface of nerve cells and excite nerve cells to death. It has been recently shown that certain drugs which block the action of glutamate and aspartate at these excitatory receptors can protect the animal brain against damage associated with stroke, cardiac arrest or perinatal asphyxia. Thus, *it is an ironic fact that today knowledgeable neuroscientists in many parts of the world are working fervently to develop methods for preventing endogenous excitotoxins from damaging the human brain, while other elements of society, including the FDA, are promoting and sanctioning the adulteration of foods with unlimited amounts of exogenous excitotoxins which are known to destroy nerve cells in the mammalian brain following oral intake."*

PHENYLALANINE

Most dietary proteins contain about four to five percent phenylalanine by weight. In sharp contrast, the concentration in aspartame may reach 56 percent (Jobe 1987). Phenylalanine tastes sweet to humans, and is preferred to water by some mammals (Ninomiya 1987).

Several aspects of phenylalanine metabolism, blood-brain concentrations, and toxicity will be briefly reviewed here because of their possible clinical relevance and for better understanding my concern.

Influence of Method of Administration

Both phenylalanine and aspartame concentrations in the blood plasma are significantly higher when aspartame is given as a solution rather than in capsule form (Chapter 4). In other words, drinking a soda containing aspartame would produce higher blood levels of the two amino acids than after taking the same amount of aspartame in powdered (capsule) form.

This fact is quite significant in interpreting studies that involve aspartame administration to normal subjects, and to patients with headaches, seizures and other complaints. For the most part, the findings are based on results after taking capsules of aspartame, as explained below.

Relevance to Provocative and Double-Blind Studies

Most of the provocative and double-blind tests performed with aspartame have entailed giving it in a capsuled dry form, or as a non-heated, aspartame-containing product within one hour of preparation (like a pudding). In my opinion, the failure to have administered aspartame products that were heated or stored for longer periods—as occurs in "real life"—mars the accuracy and significance of such studies. I have pleaded for the repetition of double-blind studies "using commercial aspartame-containing products prepared and consumed under actual conditions of use" (Roberts 1988e).

This reservation has serious clinical and public-health implications. Aspartame-associated seizures or behavioral changes in children that were dismissed as "merely anecdotal" on the basis of "negative" conclusions from such "scientific" double-blind tests. For example,

> Kreusi and colleagues (1986) studied aggression and activity in 30 preschool boys after ingesting sugar or aspartame. They concluded that neither sugar nor aspartame significantly disrupted behavior. The drinks were served cold, however, and the sweeteners added as syrups that had been kept refrigerated and in crushed ice (personal communication, February 9, 1987).

Phenylalanine Hydroxylase

Under normal conditions, phenylalanine is broken down by the body to another amino acid known as tyrosine. An enzyme called phenylalanine hydroxylase is necessary to accomplish this conversion.

If there is a lack of this enzyme, as in the inherited disorder called *phenylketonuria* (PKU), phenylalanine cannot be metabolized. It then accumulates in the blood and tissues. High concentrations of phenylalanine can cause severe *mental retardation* in children. In fact, about 1% of patients in mental institutions have this condition.

The point being stressed here is that in this and certain other circumstances aspartame products can elevate blood phenylalanine levels significantly, albeit not as high as in PKU. Furthermore, a high intake of phenylalanine tends to depress the activity of the enzyme phenylalanine hydroxylase (Harper 1970).

- High phenylalanine concentrations are found in patients with chronic kidney failure (uremia), attributed to impairment of conversion (Furst 1980).
- Phenylalanine levels may increase in the absence of sufficient insulin, which stimulates the synthesis of phenylalanine hydroxylase (Tourian 1975). This clearly could be significant for some diabetic patients (see Chapter 25).

When analyzing experimental data, it also is necessary to be aware of the marked *species differences* that exist in phenylalanine hydroxylation. For example, this process is about five times slower in man than rodents.

Other Patients Prone to Elevated Phenylalanine Levels

Increased plasma phenylalanine levels have been found in a number of other conditions and medical disorders. They include iron deficiency (Lehmann 1986), pregnancy (Yakymyshyn 1972), the use of oral contraceptives (Landau 1967, Craft 1971), cirrhosis of the liver (Jagenberg 1977, Heberer 1980, Dhont 1982), malnutrition (Antener 1981), obesity (Brown 1973, Caballero 1987), kidney disease (Pickford 1973, Jones 1978, Furst 1980, Stonier 1984), infection (Wannemacher 1976), and leukemia (Wang 1961).

Because patients with these conditions already may have high phenylalanine levels, it is understandable that adding even more of this amino acid by eating or drinking aspartame products might increase their risk for aspartame toxicity and clinical reactions.

Then, too, persons with these various conditions—characterized by impaired phenylalanine metabolism and elevated blood levels of phenylalanine—could readily be *misdiagnosed as PKU carriers.*

The "bottom line": there are a substantial number of people in whom the combination of an underlying disorder and modest or high aspartame intake are subject to the consequences of too much phenylalanine.

Phenylalanine Loading

Plasma phenylalanine concentrations increase as much as ten times when normal adults and PKU carriers are given loads of 100–200 mg aspartame/kg body weight (Matalon 1987). Using 6 mg percent phenylalanine as the blood level that is probably harmful to the outcome of a pregnancy, 14 percent of normal persons and 35 percent of PKU carriers exceed this concentration after such excessive amounts enter the body (Matalon 1987).

Phenylalanine tolerance testing is used to demonstrate a reduced capacity for metabolizing phenylalanine. Subjects ingest a single dose of this amino acid, such as 100 mg/kg body weight (Koch 1982).

Brain Levels of Phenylalanine

Consuming phenylalanine as dietary protein does not significantly elevate phenylalanine concentrations in the brain. Paradoxically, the eating of customary proteins even tends to decrease brain phenylalanine due to the concomitant increase in plasma concentrations of other neutral amino acids—especially valine, leucine, isoleucine, tyrosine and tryptophan. These amino acids compete with phenylalanine for brain uptake in the lining cells of the small blood vessels that make up the blood-brain barrier.

By contrast, ingesting an aspartame-containing soft drink with foods rich in carbohydrate and poor in protein (such as desserts) can increase—perhaps double—phenylalanine in the brain (Yokogoshi 1984).

The very high affinity for phenylalanine of the system that transports amino acids into brain capillaries (Pardridge 1987b) is emphasized. Indeed, phenylalanine has the highest affinity for transport across the blood-brain barrier of *all* the circulating amino acids. Its selective increase then can alter brain concentrations of the amino acid-derived neurotransmitters (Choi 1986).

Altered Neurotransmitter Function

Neurotransmitters are chemicals that influence brain, nerve, and endocrine function. The major neurotransmitters derived from amino acids are dopamine, norepinephrine, epinephrine, serotonin and acetylcholine.

Flooding the brain with phenylalanine, the precursor of dopamine and norepinephrine, can radically affect brain neurotransmitters. The marked rise of brain phenylalanine following aspartame ingestion—with subsequent modification of norepinephrine, epinephrine and serotonin synthesis—also reflects a reduction of neutral amino acids (Wurtman 1985). An excess of certain neurotransmitters (for example, glutamate) is known to overstimulate certain populations of nerve cells in the brain . . . even causing their destruction.

Some Objective Manifestations

Too much phenylalanine tends to "handicap the nutritional flow of amino acids to the brain" (Christensen 1987). Clinical expressions of this condition include a lowering of the seizure (convulsion) threshold in susceptible individuals, and altered mechanisms regulating hunger and satiety (Blundell 1986).

The FDA indicated its apprehension over this matter when it solicited studies for the project, "Dietary Amino Acid and Brain Function" (No. 223-86-2095, dated August 6, 1986):

> "Whether consumed as peptides or amino acids, these compounds are metabolically indistinguishable from other dietary amino acids, yet any change in the balance of amino acids, especially when large doses are consumed, has the potential to affect neurochemical processes . . . hence, a fundamental question has arisen as to whether dietary exposure to such amino acids can modulate central neurotransmitter systems and, if so, could this result in significant changes brain function and behavior . . . And at this time, there is insufficient knowledge regarding diet, neurotransmitters, and brain function."

Neurotoxicity In Infancy

Elevated phenylalanine concentrations are particularly undesirable during pregnancy and infancy because the potential exists for damaging the brain during its period of maximum growth. For example, high concentrations of phenylalanine tend to inhibit the turnover of myelin that protects nerves (Hommes 1987).

Dr. William M. Pardridge (1986) considers phenylalanine to be the major neurotoxic amino acid of aspartame. He challenged the assumption by the Council on Scientific Affairs of the American Medical Association (1985) that only one percent of the population—including children—consumes more than 34 mg aspartame/kg daily.

A five-fold increase in the plasma phenylalanine concentration (from 50 to 250 umol/L) could impair brain function in the developing fetus and in children. The most convincing evidence for such severe impairment is found among infants afflicted with phenylketonuria (PKU). Brain function deterioration associated with increased concentrations of phenylalanine and other amino acids also occurs in patients with liver failure (Fischer 1971).

Other Sources of Pure Phenylalanine

Considerable phenylalanine in forms other than aspartame is being consumed by the public as over-the-counter (OTC) products widely promoted for the relief of arthritis, migraine, whiplash, depression, dyslexia, alcoholism and obesity. Several aspartame reactors experienced worse reactions after using OTC preparations containing L-phenylalanine and DL-phenylalanine than with aspartame.

Scientists and regulatory agencies in the United States and Canada have decried such easy availability of pure phenylalanine and other amino acids because of their potential for adversely affecting brain and neurotransmitter function. Dr. C. Wayne Callaway (1986) emphasized that many physicians do not appreciate " . . . the potentially harmful effects of supplementation with megadoses of single amino acids or other nutrients."

Other Possible Pathologic Changes

Pathologic changes induced by excessive amino acid administration have been termed "amino acid imbalance," "amino acid antagonism," and "amino acid toxicity" (Brown 1967).

Excessive phenylalanine may interfere with the ability of tissues other than the brain to metabolize amino acids.

- This is most impressively demonstrated in children with PKU.
- After four weeks of phenylalanine administration, rats develop weight loss, marked pancreatic and liver changes, thickening of the skin (hyperkeratosis), and inhibition of sperm production (Klavins 1967).

ASPARTIC ACID

Aspartic acid (see Figure 4-1) constitutes about 40 percent of the aspartame molecule. This amino acid is involved in major metabolic processes. As noted earlier, it does not cross the blood-brain barrier as readily as phenylalanine.

The rise of plasma aspartic acid is decreased when it is eaten with carbohydrate or protein. Conversely, the ingestion of considerable aspartic acid, as in aspartame–containing "diet" beverages, is likely to result in higher concentrations.

The ramifications of increased aspartic acid intake require more attention than heretofore given. These considerations are germane:

- Aspartate and glutamate (two amino acid neurotransmitters) account for up to 30 percent of the total free amino acid content of the brain.
- Since severe confusion and memory loss are frequent among aspartame reactors (Chapter 11), the markedly diminished aspartate binding in the brain cells of patients with Alzheimer's disease (Procter 1986) is noteworthy.
- Aspartic acid and glutamic acid (glutamate) not only have similar chemical structures and transport systems, but also comparable toxicity in animals. Considerable information already exists concerning the toxicity of monosodium glutamate (MSG).
- D–aspartic acid, the uncommon stereoisomer, tends to accumulate in the aging human brain (Man 1983). This could have considerable significance in Alzheimer's disease and related disorders (Chapter 28).
- Furthermore, D–aspartate binding is reduced in the cerebral cortex of patients with Alzheimer-type dementia (Cross 1987).

There are other evidences for the neurotoxicity of excess aspartic acid, especially in conjunction with glutamate and MSG in the young (Olney 1970, 1979, 1980; Finkelstein 1983, Stegink 1984). Some infants lack the enzyme required to metabolize aspartic acid (Chapter 24). Moreover, the neurotoxic effects of aspartate and glutamate appear to be additive.

Other Potentially Toxic Breakdown Products

Breakdown products of aspartame, other than its two amino acids and methanol, may also contribute to adverse reactions from aspartame products. Their nature and biological activity, however, are still not fully understood for this "most studied product in history"—one that remains classified "generally regarded as safe" (GRAS). The FDA suggested that inability to account for 30 percent of aspartame decomposition products was largely explainable by the performance of analyses at temperatures of 104°F (40°C) rather than at lower temperatures (*Federal Register* February 22, 1984, p.6675).

Dr. Jeffrey Bada, a University of California (San Diego) professor of chemistry and researcher at the Amino Acid Dating Laboratory of the Scripps Institution of Oceanography, has made significant contributions in this field. He detected up ten different breakdown products of aspartame (personal communication, November 1986).

Since the definitive effects of these degradation products have not yet been determined, it seems foolhardy to dismiss them arbitrarily. Paradoxically, producers may add more aspartame to certain soft drinks in an effort to compensate for changes in sweetness resulting from the presence of these breakdown products (metabolites).

The many alterations in the physical state of aspartame, phenylalanine and aspartic acid that result from heating, prolonged storage, and interactions with other chemicals cannot be detailed here. Several that are highly relevant to my present concern, however, will be briefly mentioned.

The Effects of Heat and Storage

There is greater breakdown (degradation) of the aspartame in "diet" drinks following exposure to temperatures in excess of those used for company-sponsored studies.

- The *Federal Register* (Volume 48, No. 1322, July 8, 1983) qualified the spoilage of aspartame as 38 percent at 86°F, and over 50 percent at 104°F. (The latter is still far lower than the temperatures of some enclosed trucks in which such products are transported.)
- Boiling causes an internal rearrangement of aspartame—the L-isomers of phenylalanine and aspartic acid being changed to their unnatural D-isomers.

Interactions With Chemicals and Foods

The amino acids of aspartame may be converted to their D and L forms at room temperature from interactions with other components of food and beverages, such as chocolate (Novick 1985).

Hussein et al (1984) demonstrated that aspartame can react with certain aldehydes used as flavor compounds.

Dr. Bada found differences in the reactions of a popular aspartame tabletop sweetener and pure aspartame after both were heated. This possibly reflects the former's higher aldehyde content (personal communication, January 1987).

Questions About Diketopiperazine

A diketopiperazine (DKP) derivative of aspartame is formed in solution and during storage, especially at higher temperatures (Boehm 1984). This toxic metabolite is generally absent from the human's diet. Kulczycki (1986, 1987) suggested that DKP might act as an intermediate antigen in patients with allergic reactions to aspartame (Chapter 18).

CHAPTER SIX

Methyl Alcohol

*It was all very well to say "Drink me," but the
wise little Alice was not going to do that in a
hurry. "No, I'll look first," she said, "and see
whether it's marked 'poison' or not;" . . . she had
never forgotten that if you drink very much from
a bottle marked "poison," it is almost certain to
disagree with you, sooner or later.*
— *LEWIS CARROLL*
(*Alice's Adventures in Wonderland*)

METHYL ALCOHOL (METHANOL, WOOD alcohol), the
third breakdown product of aspartame, is a metabolic poison. It can
cause serious tissue damage—especially blindness—and even death.

Rarely found in nature as its "free" form, methyl alcohol is usually
derived or produced from other substances. It made headlines recently when 25 persons in Italy died after drinking table wine containing 5.7 percent methanol. In one series of patients with methyl
alcohol poisoning (Bennett 1952), the lowest fatal dose was three
teaspoons of 40 percent methanol.

As a known poison, it is logical to consider the possible role of
methyl alcohol in reactions to aspartame-containing products. This
applies especially to eye and neuropsychiatric complaints.

QUANTITATIVE CONSIDERATIONS

Taking aspartame into the body yields approximately *ten percent* methanol by weight (*Federal Register* February 22, 1984, pp.6672–6682). The actual methanol composition of the aspartame molecule is 32/394, or 10.9 percent.

The following approximations are provided for comparative purposes:

- 19 gm aspartame, the equivalent of one teaspoon sugar, yields 1.9 mg methanol.
- One liter of most aspartame-sweetened soft drinks contains about 55 mg methanol.
- Methanol concentrations in aspartame-sweetened beverages increase with heating and during prolonged storage.
- The amount of methanol ingested by heavy consumers of aspartame products could readily exceed 250 mg daily (Monte 1984). This is 32 times the limit of consumption recommended by the Environmental Protection Agency (EPA).
- "Abuse doses" (100 mg aspartame/kg body weight, or more) result in significant elevations of blood methanol concentrations in normal subjects (Stegink 1984). Moreover, the level remains detectable for *eight or more hours.*
- Monte (1984) calculated that one-hundredth the fatal level (a standard criterion for safety used by the FDA) translates into only *two* 12-ounce cans.

METABOLISM

Methyl alcohol appears to be the first component of aspartame released within the upper small intestine. Its absorption is rapid.

Man is highly vulnerable to methanol toxicity, largely because two enzymes required to metabolize it have been lost during human evolution (Roe 1982). Its *rate of oxidation* (or breakdown) is only one-seventh that of customary ethyl alcohol (ethanol).

The body attempts to *detoxify* methanol by oxidizing it to formaldehyde, then to formate or formic acid, and ultimately to carbon dioxide which is blown off in the breath. Formate itself may contribute to toxicity—most notably as metabolic acidosis and eye damage.

The *rate of methanol elimination* in humans is five times slower than for a similar amount of ethyl alcohol (Forney 1968). Accordingly, the daily ingestion of "individually innocuous amounts of methyl alcohol" could result in "eventual poisonous effects."

Interactions of Methanol With Other Drugs

Used together, the interactions between methanol and *chemical compounds related to ethyl alcohol* might have clinical significance (Posner 1975). These could include the oral (sulfonylurea) drugs used in treating diabetic patients, metronidazole (an anti-bacterial agent), and allopurinol (a standard drug used for managing gout).

The consumption of methyl alcohol in the form of aspartame products theoretically may harm alcoholic patients being maintained on *disulfiram*. Antabuse®, the trade-name drug, is currently being taken by an estimated 400,000 persons in the United States, while at least as many use less expensive generic brands.

Methanol and formaldehyde concentrations could rise in patients receiving maintenance disulfiram who are excessive consumers of aspartame products due to (1) a further slowing of methyl alcohol degradation, and (2) inhibition of the enzyme aldehyde dehydrogenase.

- Koivusalo (1958) reported that the rate of methanol elimination in rabbits was considerably delayed by disulfiram. Methanol was still detectable in the blood 100 hours after the smallest administered dose.
- Way and Hausman (1950) noted more rapid toxicity from oral methanol in rats and rabbits after prior disulfiram administration.

THE METHYL ALCOHOL SYNDROME

The symptoms and signs of methanol toxicity in man are diverse. Dr. Woodrow Monte (1984), Director of the Food Science and Nutrition Laboratory at Arizona State University, reviewed *the methyl alcohol syndrome*. Its symptoms and signs need *not* correlate with blood concentrations of methanol.

The disorders caused by methyl alcohol are listed below because of their possible relevance to complaints encountered among certain reactors to aspartame products. Admittedly, such relationships, as well as the concept of methanol as a "cumulative poison," have been denied or criticized by representatives of the manufacturer (Sturtevant 1985).

Eye Damage

Blindness caused by methanol has been attributed to the toxic effects of its breakdown products, formaldehyde and/or formic acid, on retinal cells. Methanol produces swelling of the optic disc (Hayreh 1977) and degeneration of retinal ganglion cells (Baumbach 1977) in monkeys. This topic is further discussed in Chapter 14.

Brain Involvement

CT scans of the brain in patients with methanol poisoning have revealed areas of presumed local death or infarction (McLean 1980, Swartz 1981). Accordingly, a review of some observations concerning brain swelling (edema), slowing of blood in its vessels (vascular stasis), and altered cerebral function after experimental methanol exposure is germane.

- A marked reduction of both cerebral blood flow and cerebral oxygen consumption has been documented during methanol poisoning.
- Brain swelling occurs in both humans and experimental animals (Menne 1938, Bennett 1953, Erlanson 1965, Rao 1977).
- Significant alterations of brain water, sodium and potassium, with concomitant vascular stasis, are found after methanol administration—both acute and chronic—in male rabbits and monkeys (Rao 1977). Furthermore, the progressive rise of blood methanol levels after the third week suggests partial inhibition of methanol degradation.
- Survivors of severe methanol intoxication have developed Parkinsonism, dementia and other neurologic abnormalities . . . as well as blindness (McLean 1980).

Involvement of the Peripheral Nerves (Neuropathy)

Neuropathic symptoms from methanol include numbness, "pins and needles" sensations (paresthesias), and shooting pains (Chapter 12). They are particularly evident after chronic exposure to methanol.

Inflammation of the Pancreas (Pancreatitis)

Pancreatitis has been reported in methanol poisoning (Bennett 1952). Pancreatitis may have produced the severe abdominal pain in some aspartame reactors in the present series (*see* Chapter 19).

Inflammation of the Heart Muscle (Cardiomyopathy)

Cardiac changes have been found in patients with methanol poisoning. The relatively frequent complaints of palpitations, rapid heart action and atypical chest pain among aspartame reactors (Chapter 20) may be pertinent.

Metabolic Acidosis

Methanol characteristically causes metabolic acidosis. In this severe biochemical state, excessive acids in the body can result in respiratory failure and death. (Diabetic acidosis is another type of metabolic acidosis.)

Clinical acidosis might be induced after large amounts of aspartame are ingested, especially by children. A London physician wrote me about his 9-year-old daughter who had been given an aspartame-containing cola for recurrent abdominal pain.

". . . in all, she probably drank about 1.5 litres over a 24-hour period together with eating a few slices of toast in the same 24 hours. The following morning she was found semi-conscious, confused, and had a metabolic ketoacidosis but a normal blood sugar on admission to hospital. Tests for metabolic poisons such as aspirin were negative. Fortunately, rehydration restored her to normal within 4–5 hours biochemically together with restoration to normal levels of consciousness. I thought at the time that (the diet cola) could be responsible but could not find any evidence to support this . . . the (company) in the UK (was) very defensive about the effects of aspartame and denied any knowledge of adverse effects of aspartame."

METHANOL IN "NATURAL SOURCES"

The FDA and the Council on Scientific Affairs of the American

Medical Association (1985) rationalize the safety of methanol in aspartame products with statements such as "fruits and vegetables are also sources of dietary methanol," and "dietary methanol also arises from fresh fruits and vegetables" (*Federal Register* Vol. 48, No. 132, July 8, 1983, p. 31380). The *FDA Talk Paper* (January 24, 1984) further asserted: "FDA said that no safety issues appeared to be involved, there being more methanol in many fruit juices than in long-stored aspartame products, including carbonated beverages." Yet, Monte (1984) estimated the averaged daily intake of methyl alcohol from natural sources at less than 10 mg.

In giving final approval for the addition of aspartame to carbonated beverages and carbonated beverage syrup bases (*Federal Register* February 22, 1984, pp. 6672-6682), the FDA stated:

> The agency does not believe that methanol exposure equivalent to 10 percent of the aspartame dose is of sufficient quantity to be of toxicological concern under acute or chronic use conditions . . . FDA remains convinced that the studies submitted by Searle in support of the dry use, and reviewed by the FDA prior to the dry uses approval and again its evaluation of the carbonated beverage petition, adequately support the agency's conclusion that there was "no cause for concern from the levels of dietary methanol resulting from the highest projected levels of aspartame consumption (48 FR 31376 at 33181)."

After many requests, I still have been unable to obtain data from the FDA concerning *its* assays for the free methanol content of fruits, fruit juices, alcoholic beverages and other products derived by current methods.

- The reference used by the FDA concerning the presence of more methanol in the "average fruit juice" (Francot 1956) than aspartame-containing orange soda was published three decades earlier.
- Dr. David G. Hattan (Chief, Regulatory Affairs Staff, Office of Nutrition and Food Sciences, Center for Food Safety and Applied Nutrition of the FDA) told me that he was not aware of any such recent analyses (personal communication, June 8, 1987).

Three older references on this subject are repeatedly cited. They were published in French (Le Moan 1956), German (Sommer 1962), and Russian (Ivanitskiy 1973). I question their relevance to aspartame, however, owing to the emphasis on methanol in pectin-containing fruit and fruit products, some vegetables, wines and other alcoholic beverages. In fact, Ivanitskiy (1973) concluded: "From the standpoint of health, it is not correct to apply the standards of the methanol content of alcoholic beverages arbitrarily to fruit juices, as has been done by several authors."

Another issue must be raised in this context. Any effects of small amounts of methanol possibly present in fruit juices and wine tend to be offset by their caloric content and the presence of ethanol (Gilger 1959). This contrasts with the absence of nourishment in the case of "diet" drinks.

Recent studies by Lund et al (1981) and Nisperos-Carriedo and Shaw (1989) have clarified the methanol, ethanol, and acetylaldehyde concentrations of citrus products. Contemporary gas chromatography methods were used. These data indicate the following:

- Orange juice and grapefruit juice average as much as ten times more ethanol, and from two to ten times more acetylaldehyde, than methanol.
- The concentration of methanol is higher in fresh-squeezed orange juice compared to the small amounts in pasteurized orange juice (22mg/liter), frozen concentrate (3.4mg/liter), reconstituted juice from concentrate (trace/one glass), and orange juice in tin cans (trace). The former is undoubtedly due to persistence of pectinmethylesterase enzyme in the unpasteurized juice, which demethylates some pectin and liberates methanol in the process.

A Clinical Overview

*If the clinician, as observer, wishes to see things as
they really are, he must proceed without any
preconceived notions whatsoever.*
— DR. J. M. CHARCOT

JUST WHAT SIGNS AND symptoms can aspartame products
cause? Analysis of the data on the first 551 persons in my registry
who appeared to have adverse reactions to aspartame-containing
products provides a perspective. The group consisted of 160 individuals in my own files (consisting of 55 patients seen in consultation,
20 persons who were interviewed in person or by phone, and 85
who corresponded with me directly after hearing or learning of my
interest in aspartame), and 351 consumers who had first contacted
various cooperating groups concerned with aspartame reactions.
These included Aspartame Victims And Their Friends (295 persons),
the Community Nutrition Institute (68 persons), and Dr. Woodrow
Monte of Arizona State University (28 persons).

The most common complaints ascribed to aspartame products
were

- severe headache
- seizures (convulsions)
- impairment of vision
- dizziness
- atypical unexplained pain involving various sites (the eyes,
  ears, face, neck or chest)

- rashes
- extreme fatigue
- depression
- a change of personality
- confusion and memory loss

CRITERIA FOR INCLUSION IN STUDY

Several essential criteria had to be met before categorizing an individual as an "aspartame reactor." The main requirements were (1) the *prompt* improvement of most symptoms and signs after stopping and avoiding aspartame-containing products, and (2) their *predictable* recurrence within hours or days following reuse or rechallenge (either deliberately or inadvertently), or during provocative testing by the person or an investigator. In point of fact, many had retested themselves more than ten times "just to be certain." Nearly two-thirds of correspondents who completed the questionnaire indicated that they felt better within *two days* after stopping the use of aspartame products.

Another general criterion used for inclusion in this series was the impression that these individuals (or other persons providing the essential information) seemed reasonable and intelligent. This premise assumes significance when one is dealing with patients and consumers who offer several complaints that may not be accompanied by overt signs or laboratory findings—e.g., headache, itching without a rash, severe confusion, unexplained chest pain.

Astute physicians have attempted to convey their perceived intelligence of patients when publishing on unusual clinical phenomena, partly in anticipation of criticism about subjective complaints. For example, J.M.S. Pearce (1988), a neurologist, described his first case of the "exploding head syndrome" in *The Lancet* as "a sensible woman."

THE SURVEY QUESTIONNAIRE

I devised a survey questionnaire to obtain more comprehensive information about persons who believed they had adverse reactions to aspartame-containing products, and the nature of their complaints. These data were computerized and analyzed with the assistance of Data Information Services at the Good Samaritan Hospital (West Palm Beach).

The questionnaire is reproduced in its entirety at the end of this chapter. Repeated reference will be made to it throughout the book. Also, study of this form may prove a valuable exercise for readers who suspect they are aspartame reactors . . . as was the case with many of my patients and correspondents.

> A 57-year-old engineer-executive presented with multiple complaints. He felt better within two weeks after discontinuing aspartame products. After receiving the survey questionnaire, he wrote:

> "I was intrigued by the questionnaire since it holds hope for me that the wide array of seemingly unrelated symptoms could come from a common causative factor."

Since many conclusions of this study are influenced by the results of this questionnaire, it is fair to ask if this type of information is valid or just "anecdotal." Dr. Ephraim Kahn (1987), a specialist in environmental epidemiology, appropriately answered this question in another context.

> "But reports by patients or parents in response to a pre-designed questionnaire and as part of a study design, although subjective, are not anecdotal. Their validity would depend on the quality of the study design and the care with which the study was performed."*

SYMPTOMS AND SIGNS

The major symptoms and signs experienced by 551 aspartame reactors are listed in Table 7-1. The most common problems will be discussed individually in ensuing chapters.

* © 1978 *Journal of Pesticide Reform*. Reproduced with permission.

Table 7-1. Complaints in 551 Aspartame Reactors (Rounded
Percentages)

Eye
Decreased vision and/or other eye problems (blurring,
"bright flashes," tunnel vision)   140   (25%)
Pain (one or both eyes)   51   (9%)
Decreased tears, trouble with contact lens, or both   46   (8%)
Blindness (one or both eyes)   14   (3%)

Ear
Tinnitus ("ringing," "buzzing")   73   (13%)
Severe intolerance for noise   47   (9%)
Marked impairment of hearing   25   (5%)

Neurologic
Headaches   249   (45%)
Dizziness, unsteadiness, or both   217   (39%)
Confusion, memory loss, or both   157   (29%)
Severe drowsiness and sleepiness   93   (17%)
Paresthesias ("pins and needles", "tingling") or numb-
ness of the limbs   82   (15%)
Convulsions (grand mal epileptic attacks)   80   (15%)
Petit mal attacks and "absences"   18   (3%)
Severe slurring of speech   64   (12%)
Severe tremors   51   (9%)
Severe "hyperactivity" and "restless legs"   43   (8%)
Atypical facial pain   38   (7%)

Psychologic-Psychiatric
Severe depression   139   (25%)
"Extreme irritability"   125   (23%)
"Severe anxiety attacks"   105   (19%)
"Marked personality changes"   88   (16%)
Recent "severe insomnia"   76   (14%)
"Severe aggravation of phobias"   41   (7%)

Chest
Palpitations, tachycardia (rapid heart action), or both   88   (16%)
"Shortness of breath"   54   (10%)
Atypical chest pain   44   (8%)

Table 7-1. *Complaints in 551 Aspartame Reactors (Rounded Percentages)—continued*

Recent hypertension (high blood pressure)	34	(6%)
Gastrointestinal		
Nausea	79	(14%)
Diarrhea	70	(13%)
Associated gross blood in the stools . . . 12		
Abdominal pain	70	(3%)
Pain on swallowing	28	(5%)
Skin and Allergies		
Severe itching without a rash	44	(8%)
Severe lip and mouth reactions	29	(5%)
Urticaria (hives)	25	(5%)
Other eruptions	48	(9%)
Aggravation of respiratory allergies	10	(2%)
Endocrine and Metabolic		
Problems with diabetes: loss of control; precipitation of clinical diabetes; aggravation or simulation of diabetic complications	60	(11%)
Menstrual changes	45	(8%)
Severe reduction or cessation of periods . . . 22		
Paradoxic weight gain	34	(6%)
Marked weight loss	26	(5%)
Marked thinning or loss of the hair	32	(6%)
Aggravated hypoglycemia (low blood sugar attacks)	25	(5%)
Other		
Frequency of voiding (day and night), burning on urination (dysuria), or both	69	(13%)
Excessive thirst	65	(12%)
Severe joint pains	58	(11%)
"Bloat"	57	(10%)
Fluid retention and leg swelling	20	(4%)
Increased susceptibility to infection	7	(1%)

MULTIPLE COMPLAINTS

The majority of aspartame reactors had multiple symptoms and signs. Most of these complaints improved or disappeared when aspartame products were stopped, and recurred after resuming them.

A 34-year-old woman consumed large amounts of aspartame-containing products for 1-1/2 years . . . including eight glasses of a hot chocolate *daily*. She temporarily lost her job as a receptionist when disabled by the following complaints:

- A marked decrease of vision and pain in both eyes
- Ringing in the left ear
- Sensitivity of noise in both ears
- Headaches
- Dizziness and lightheadedness
- Cramping of the toes and legs every day
- Two convulsions
- Memory loss ("I couldn't remember something that was said recently")
- Tingling, pins and needles, and numbness of the arms and legs
- Severe depression
- Extreme irritability
- Anxiety attacks
- Marked "personality changes"
- Fear of crowds and other phobias ("I felt I wanted to ruin someone who ruined me")
- Pain in the abdomen
- Nausea and vomiting
- Diarrhea with bloody stools
- Marked abdominal bloat
- Considerable thinning of the hair
- A weight gain of 10 pounds
- Dryness of the mouth
- Intense thirst
- Frequency and burning of urination, with an associated urinary-tract infection

Four physicians and consultants were seen, and she was hospitalized "for observation" . . . but to no avail. Multiple studies—including X-rays of the head, a CT scan of the brain, an EEG, two upper gastrointestinal series, two barium enemas, allergy tests, and special examinations of the eyes and ears—proved inconclusive. The patient *herself* then deduced that aspartame had been causing or aggravating these complaints, and felt improved within three days after avoiding these products. She retested herself *more than ten* times . . . on each occasion promptly experiencing a recurrence of symptoms (especially vomiting).

A 52-year-old woman had many complaints over two weeks while consuming three cans of an aspartame soft drink, four packets of a tabletop sweetener, three glasses of presweetened ice tea, and one cup of aspartame-sweetened hot chocolate daily. She wrote:

"After experiencing slurred speech, severe headaches, poor coordination and depression, my adult son suggested a possible aspartame reaction. I stopped it, thinking that maybe it was psychological. I tried it again two times, and the symptoms returned!"

Some reactors believed they had only one aspartame-associated symptom. They realized otherwise, however, when interviewed or while completing the questionnaire.

A 38-year-old-woman felt that her reaction to aspartame was "purely intestinal in nature"—namely, severe abdominal pain, nausea, bloody diarrhea, and abdominal bloat. But she also complained of recent "dry eyes" with difficulty in wearing contact lens, and had gained 15 pounds—other common reactions to aspartame products.

FDA DATA ON ASPARTAME COMPLAINTS

The data of the FDA on 3,326 complaints *volunteered* by consumers who ascribed their problems to aspartame products (Tollefson 1987) appear in Table 7-2.

Table 7-2.   *Complaints Reported to the FDA by 3,326 Aspartame
                  Complainants* (Rounded Percentages)*

Headache	951	(19%)
Dizziness or problems with) balance	419	(9%)
Change in mood quality or level	349	(7%)
Vomiting and nausea	329	(7%)
Abdominal pain and cramps	254	(5%)
Diarrhea	178	(4%)
Change in vision	162	(3%)
Fatigue, weakness	141	(3%)
Seizures and convulsions	137	(3%)
Sleep problems	127	(3%)
Memory loss	125	(3%)
Rash	111	(2%)
Change in sensation (numbness, tingling)	91	(2%)
Hives	80	(2%)
Other	1464	(30%)

*From Tollefson, L., Barnard, R. J., Glinsmann, W. H.: Monitoring of
adverse reactions to aspartame reported to the U.S. Food and Drug Administration. In *Proceedings of the First International Conference on Dietary Phenylalanine and Brain Function,* Edited by R. J. Wurtman and E. Ritter-Walker,
Washington, D.C., May 8–10, 1987, pp.347–372. Reproduced with permission of Dr. R. J. Wurtman.

THE SURVEY QUESTIONNAIRE

Palm Beach Institute for Medical Research, Inc.
300 27th Street
West Palm Beach, Florida 33407

H.J. Roberts, M.D.
Medical Director

VOLUNTARY QUESTIONNAIRE
FOR PERSONS WITH POSSIBLE REACTIONS
TO PRODUCTS CONTAINING ASPARTAME

\# _____

I. GENERAL INFORMATION

NAME (Omit if You Wish) _____

ADDRESS (Omit if You Wish) _____

_____

TELEPHONE NUMBER (Omit if You Wish) Area Code   \#

_____

SEX (Circle)   Male   Female

RACE (Circle)   White   Black   Asian_____   Other_____

PRESENT AGE (years)_____
OCCUPATION _____

II. INFORMATION ABOUT YOUR PRIOR USE
OF ASPARTAME

Age when your first reaction occurred — _____ years

How long had you been eating or drinking aspartame before then?

(Note: Aspartame was first put in soft drinks in July 1983).

_____ Days _____ Weeks _____ Months _____ Years

_____

_____

Did you *greatly* increase such use *before* these symptoms occurred?
(Circle)   Yes   No

(If "Yes," how long *before* your symptoms occurred?)
_____ Days _____ Weeks _____ Months _____ Years

Why did you use or increase products sweetened with aspartame? (Circle)

   I preferred their taste
   Great thirst in hot weather
   I wanted to avoid sugar because of: (circle)
     Diabetes
     Hypoglycemia ("low blood sugar")
     An overweight problem
     A skin condition
     Other _____

Did you use aspartame-sweetened products under these conditions? (Circle and insert for how long)
While *pregnant* for   _____ weeks   _____ months
While *breast-feeding* for   _____ weeks   _____ months

How much do you estimate that you were eating or drinking aspartame-sweetened products *daily* when your symptoms began?

Some examples: diet cola beverages, diet soft drinks, various brand chocolate mixes, gelatins, puddings, pre-sweetened cereals, iced tea mixes, powdered soft drinks

_____ cans (12 ounces) of (name of product)_____
_____ small bottles (6 ounces) of (name)_____
_____ large bottles (liter or 33 ounces) of (name)_____
_____ very large bottles (2 liters or 67 ounces) of
     (name)_____
_____ packets of table-top sweetener (name)_____
_____ regular glasses of iced tea (name)_____
_____ regular glasses of soft drink mixes (name)_____
_____ bowls of cereal (name) _____
_____ servings of puddings or gelatins (name)_____
_____ sticks of gum (name)_____
_____ (envelopes) or (teaspoons) of other products
     (name) _____

Were you on a *strict* diet to lose weight when your problem began?   (Circle)   Yes   No

If "Yes," for how long? _____weeks _____months

Were you *really* exercising to lose weight when your problem began?   (Circle)   Yes   No

III. SUSPICIONS ABOUT ASPARTAME CAUSING
OR AGGRAVATING YOUR CONDITION

Who or what *first* made you think that aspartame might be causing your symptoms? (Circle)

Myself
A relative
A friend
My own doctor
A nurse
A newspaper article
A TV program
A medical specialist (name) _____
Other (please state) _____

Did you then *completely* stop using aspartame?
(Circle)   Yes   No

If you stopped, *did your symptoms improve?*
(Circle)   Yes   No

How soon did your symptoms *begin* to improve *after* you stopped?

_____ Days _____ Weeks _____ Months

Did *all* of your symptoms *disappear* after you stopped?
(Circle)   Yes   No

If "Yes," how soon? _____ Days _____ Weeks _____ Months

If you felt better, what problems *still* persisted?

_____

_____

If you felt better, did your symptoms *return* when you ate or drank these products *again*?   (Circle)   Yes   No

If "yes," *how soon* did the symptoms come back?
_____ Days _____ Weeks _____ Months

How many times did you then *retest* yourself before you became *convinced* that aspartame product(s) *really* caused your symptoms?
(Circle)   0   1   2   3   4

Did the symptoms return after you ate or drank something *you didn't know* contained aspartame? (Circle)   Yes   No

If "Yes," what was the product? _____

Did your symptoms *significantly* affect your business and personal life?
(Circle) Yes   No

If "Yes," how or how much? (Circle)
   A bit
   Severely (loss of job, stopped school, divorce, etc.)
   I was incapacitated
   I lost my driver's license

Describe fully _____

_____

IV. REACTIONS I ATTRIBUTE TO ASPARTAME

Please circle *any* of the following complaints that you or your doctors thought might be caused by, or aggravated by, aspartame. (Elaborate later if you wish.)

Eyes
   A marked decrease of vision in   One eye   Both eyes
   Loss of vision in   One eye   Both eyes
   Pain in   One eye   Both eyes
   Recent "dry eyes"   Yes   No
   Recent trouble wearing contact lens   Yes   No

Ears
   Ringing or "cracking" in   One ear   Both ears
   Loss of hearing in   One ear   Both ears
   Severe sensitivity to noise in   One ear   Both ears

Brain and Head
   Severe headaches
   Severe dizziness or feeling lightheaded
   Severe unsteadiness of my legs
   Convulsions (seizures, epilepsy)
      How many attacks? (Circle) 1 2 3 4 More_____

   Was your brain wave test (EEG) abnormal?
                        (Circle)   Yes   No

Epilepsy-like "fits" *without* convulsions — also known as petit mal, psychomotor attacks
Severe drowsiness and falling asleep (without reason)
Severe shaking (tremors)
Severe insomnia (trouble sleeping)
Severe mental confusion
Marked memory loss
Severe hyperactivity (as jumping of the arms or "restless" legs)
Severe tingling, pins and needles, or numbness of arms and legs
Unexplained pains in or around your face
Slurring of your speech
Severe depression
Did you actually think about suicide? (Circle)   Yes   No
Extreme irritability
Severe "anxiety attacks"
Marked "personality changes"
Severe aggravation of phobias (such as fear of crowds or heights)

Chest

Attacks of shortness of breath
Palpitations (heart fluttering)
Rapid heart beat attacks
Unexplained chest pains
Recent high blood pressure

Abdomen

Pain in the belly
Severe nausea
Diarrhea
Blood in the stools
Severe bloat (swelling)

Skin and Allergies

Severe itching without a rash
Hives (water blisters)
Other rashes
A marked loss or thinning of your hair
Mouth reactions

Lips became swollen
Tongue became swollen
Pain in the mouth
Pain or trouble swallowing

A Marked and Unintended Change of Weight
Gain of_____ pounds   Loss of _____ pounds

Other Problems
Severe thirst
Marked frequency of urinating   Day   Night   Both
Severe changes in menstrual periods
    More frequent   Less frequent   They stopped
Severe joint pains
Marked swelling of the legs
"Low blood sugar" (Hypoglycemia) attacks
Poorer control of my diabetes even while I was on
    Diet   Oral drugs   Insulin

Other Suspected Reactions (Please list)

_____

_____

V. CONSULTATIONS AND TESTS
BEFORE YOU STOPPED ASPARTAME

How many *physicians and consultants* did you see *in your own community?*
(Circle)   1   2   3   4   _____ More

How many *consultants or clinics* did you visit *outside of your community?*
(Circle)   1   2   3   4   _____More

Were you referred to
    A *psychiatrist*   Yes   No
    A *psychologist*   Yes   No
    Another *"therapist"*   Yes   No

How many times were you hospitalized BEFORE aspartame was suspected
    as the cause of your symptoms? (Circle)   1   2   3   4   5

Did you have any of the following tests BEFORE aspartame was suspected as the cause? If yes, circle how many.

X-rays of the head	0	1	2	3	4	5
CAT scans of the brain	0	1	2	3	4	5
MRI (magnetic resonance) of brain	0	1	2	3		
Angiograms (X-rays of blood vessels in the brain)	0	1	2			
Lumbar puncture (spinal tap)	0	1	2			
Electroencephalograms (EEG or brain wave test)	0	1	2	3	4	5
Stress test of the heart	0	1	2			
Heart monitor (12 or 24 hrs)	0	1	2			
Upper GI series with barium (X-rays of the stomach)	0	1	2			
Barium enema (X-rays of the lower bowel)	0	1	2			
Allergy tests (skin, others)	0	1	2			
Special eye tests	0	1	2			
Special ear tests	0	1	2			

How much do you estimate that these consultations and tests cost
   You out of your own pocket? $ _____
   Your insurance carrier(s) $ _____

VI. REPORTS OF SUSPECTED REACTIONS

Did You Report Your Suspected Reaction(s) to Aspartame?
               (Circle)   Yes   No

If "Yes," to whom? (Circle)
   My doctor
   A consultant
   My family
   The manufacturer of the product
   A public-health department
   The Food and Drug Administration (FDA)
   The Centers for Disease Control
   Aspartame Victims and their Friends

Another Consumer Organization
    Name _____

Dr. H. J. Roberts
    Another interested researcher I heard or read about
        (Name) _____

If "Yes," how did you do so? (Circle)
    By letter   By phone call(s)   Other (state) _____

VII. REACTIONS IN YOUR FAMILY

Have other members of your family had problems that they attrib-
    uted to aspartame?   (Circle)   Yes   No

If "Yes," how are they related, and how many (Circle or number)
    Husband
    Wife
    Daughter(s) _____

    Son(s) _____

    Others (who?) _____

What kind of reaction did they have?
_____

_____

VIII. YOUR MEDICAL BACKGROUND

Did you have *any* of the following *BEFORE* taking aspartame... and
    if so, *how long before?* (Circle and write number of months or years
    after each condition)
    Migraine
    Other headaches
    Depression (feeling blue)
    Severe "anxiety"

Poor vision
Ear problems
Heart disease
High blood pressure
Stomach trouble
Bowel trouble
Kidney trouble
Hives
Other rashes (such as psoriasis, acne or lupus)
Parkinsonism (shaking palsy)
Allergies
   Hay fever
   Asthma
   Medicines (such as aspirin, penicillin)   (List)

_____

_____

"Low blood sugar" (hypoglycemia) attacks
Diabetes treated with
                  Diet   Pills   Insulin
A Thyroid problem
   Goiter
   Underactive thyroid (hypothyroidism)
   Overactive thyroid (hyperthyroidism)
"Arthritis"
A marked weight problem (20 pounds or more)
An alcohol problem (Circle)   Long ago   Current
A drug abuse problem (Circle)   Long ago   Current
   What drug(s)? _____

Other conditions not listed above:

_____

_____

What *other* medicines (not listed above) or vitamins were you taking
as well as aspartame when your symptoms began?

(Circle and estimate *daily* amounts)
Aspirin

Multiple vitamin
Vitamin C
Vitamin E
L-Dopa (for Parkinsonism)
Aldomet (for hypertension)
Beta blockers (as Inderal®, Corgard®, Tenormin®, Lopressor®)
   for a heart or blood pressure problem
Tranquilizers
Others _____

Do you smoke *cigarettes?* (Circle)   Yes   No
If "yes," how many *daily?* (Circle)
   Less than 1/2 pack   1/2-1 pack   More than 1 pack

Did you get a severe attack after drinking *alcohol* while taking
   aspartame?   (Circle)   Yes   No

If "Yes," how much? (Circle)   Very little   Moderate   A lot

If "Yes," what kind of drink (Circle)   Wine   Beer   Liquor

IX. OTHER USEFUL INFORMATION

Describe any *unusual aspects of your reaction(s) to aspartame:*

_____

_____

_____

Have *any of your friends* had problems with aspartame?
               (Circle)   Yes   No

How many?   (Circle)   1   2   3   4

If "Yes," what kind of problems?

_____

_____

Do you *still* use products containing Aspartame?
               (Circle)   Yes   No

If "Yes," — how much? _____

         — name(s) _____

˙If "No," are you using the following?
      Sugar   Yes   No
      Honey   Yes   No
      Saccharin   Yes   No

      Other sweetener (name) _____

Have you ever heard of the condition known as *phenylketonuria* (PKU)?   (Circle)   Yes   No

   If "Yes," do you *really* know what it is?
                           (Circle)   Yes   No

   If "Yes," has anyone in your family had it?
                           (Circle)   Yes   No

Do you believe that the aspartame content of foods and drinks
should be required *on labels by law?*
                           (Circle)   Yes   No

   If "Yes," would you study them?
                           (Circle)   Yes   No

Do you think that *Congress* should do anything about regulating
the use of aspartame?   (Circle) Yes   No
If "Yes," what?

_____

_____

THANKS FOR YOUR INTEREST AND TIME!

FEEL FREE TO CONTINUE OR EXPLAIN
ON THE REVERSE BLANK SIDE

CHAPTER EIGHT

General Aspects of Aspartame Reactions

*Observation is more than seeing; it is knowing what
you see and comprehending its significance. The
process is far more mental than photographic. True
observation implies studying the object and drawing
conclusions from what is seen.*

— CHARLES GOW

SOME GENERAL OBSERVATIONS ABOUT the 551 aspartame reactors in this series are summarized here. They encompass their age range, the preponderance of females, the latent period (time lapse) before complaints began, the severity of reactions, inadvertent rechallenge, withdrawal symptoms, and comparable reactions among close relatives.

THE AGE RANGE

The average age of aspartame reactors when symptoms began was 43 years; most were in their 20s to 50s. The age range extended from infancy to 92 (two persons).

A 2-year-old girl developed a "violent rash" after drinking an aspartame soda. The eruption, and ensuing facial swelling, recurred every time she drank an aspartame-containing beverage or chewed "diet" gum.

A 92-year-old woman remained active notwithstanding a recently broken leg that confined her to a wheelchair. She developed severe vomiting and diarrhea after consuming aspartame products. These complaints did not recur when she avoided aspartame and resumed sugar. Aware of her age, she added, "I am in my right mind."

FEMALE PREPONDERANCE

Women consistently outnumbered men by a 3:1 ratio—specifically, 74 percent females and 26 percent males. This preponderance remained consistent both among my patients and correspondents who submitted the survey questionnaire.

A comparable predisposition of females was found by the FDA based on complaints it received from 3,326 complainants (Tollefson 1987). Women comprised 77 percent; three-fourths were between 20 and 59 years.

Factors That May Influence Female Preponderance

A number of factors may explain the vulnerability of women to aspartame products. The following facts are germane.

- Iron deficiency in menstruating women (Lehman 1986), which appears to impair the conversion of phenylalanine to tyrosine (Chapter 5).
- The increased proneness of women to severe depression (Holden 1986), diabetes mellitus (National Conference on Women's Health 1986), allergies (McLain 1986), and autoimmune disorders—all of which may be affected adversely.
- The significantly greater insulin responses of non-obese healthy females to phenylalanine, whether given orally or intravenously (Shah 1986).
- The higher frequency of severe reactive hypoglycemia ("functional hyperinsulinism"; "low blood sugar attacks") in women—even among the older age groups (Morton 1950, 1953; Zeytinoglu 1969).
- The considerably greater release of the hormone prolactin in normal women (compared to normal men) by phenylalanine

in doses as low as one gram (1,000 mg), presumably due to an enhanced effect of estrogen on prolactin secretion.

- Depression of glucose oxidation in brain tissue by diethylstil-bestrol, a female hormone (Gordon 1947).
- The greater mobilization of fat from fat cells (adipocytes) in the abdominal region and other depots among women after drastic caloric restriction (Leibel 1987).
- The aggravation of aspartame reactions by fluid retention and other features characteristic of the premenstrual syndrome (PMS).
- The chronic state of hyperinsulinism during pregnancy and in women taking "the pill" (Spellacy 1966, Javier 1968, Yen 1968).
- The unique responsiveness of the immune system to self-generated chemicals (antigens) in women, perhaps explaining their greater vulnerability to such autoimmune diseases as rheumatoid arthritis (3:1 ratio) and lupus erythematosus (10:1 ratio) (*Science* 238:159, 1987).

"Fear of fat," a widespread cultural phenomenon that afflicts contemporary American women, also warrants mention. Forty-two percent of aspartame reactors who completed the survey questionnaire indicated that "an overweight problem" was their prime reason for using these products. Teenagers—and even younger girls—often couple their reduced intake of calories with excessive consumption of aspartame-containing foods or beverages due to pressure from peers and the influence of high priests of fashion.

The goals of this thinness cult not only may be unattainable, but also are fraught with danger. I have described elsewhere the serious medical complications of such arbitrarily severe caloric restriction (Roberts 1985). Someone suggested that an observation by Coco Chanel, the famous fashion designer, ought to be placed on large posters in schools, churches and medical offices: "Fashion is generated by men who actually hate women."

There is an ironic aspect to the considerable use of aspartame products by weight-conscious women: paradoxic weight gain. This subject will be discussed in Chapter 17.

ASPARTAME CONSUMPTION

The 397 persons who completed the survey questionnaire provided detailed information concerning their consumption of aspartame products. The data are summarized in Table 8-1. It also includes the approximate aspartame content of each product, based on information supplied by the FDA.

LATENT PERIOD

The interval between use of aspartame products and the beginning of complaints varied not only among individuals, but also in the same person at different times. The initial period usually was several weeks or months. In contrast, many who improved after avoiding such products suffered a severe recurrence within hours or days after resuming aspartame. This sequence is reminiscent of allergic reactions to drugs once patients become sensitized to them.

A 61-year-old computer operator experienced visual difficulty, headache, memory loss, dizziness, facial pains and irritability after taking aspartame.

"It seemed to take several weeks to build up the reaction the first time, but when I tested it a month or so after I stopped its use, I got the same feelings after just a day or two of use. And one time someone gave me a glass of a soft drink that I didn't know contained aspartame. I had the same symptoms within one hour or less after the drink."

However, some individuals experienced an unequivocal severe reaction within minutes to several days following their apparent *first* exposure to an aspartame product. The most notable complaints in such persons were dizziness, swelling of the lips, itching, hives and other rashes. They offered variations of the same expression: "I knew almost immediately that aspartame was the cause."

A 60-year-old woman drank two cans of an aspartame soft drink for two days. She then developed severe lightheadedness, slurred speech, decreased vision in both eyes, and ringing in one ear.

"I knew right away on my own it was aspartame. I was in 100% perfect health before I took aspartame. I got sick. Stopped it. Feel great now."

Table 8-1. *Estimated Consumption of Aspartame-Containing Products by 397 Respondents When Adverse Symptoms Began*

Product	Size and Estimated Aspartame Content	Users	Mean of Users (Daily)	Percentage of Total Consumers*
Cola soft drinks	12 ounce can	192	2	61%
(16 mg/ounce,	6 ounce bottle	12	2	
550 mg/liter)	1 liter bottle	24	1	
	liter bottle	14	1	
Other soft drinks	12 ounce can	50	2	18%
(16 mg/ounce,	6 ounce bottle	6	2	
550 mg/liter)	1 liter bottle	12	1	
	2 liter bottle	5	1	
Tabletop sweetener	Packet (35 mg)	171	3	43%
Presweetened ice tea	Glass (100 + mg)	61	2	15%
Presweetened drink mixes	Glass (100 + mg)	102	3	26%
Presweetened hot chocolate	Glass (100 + mg)	52	1	13%
Presweetened cereal	Cup (60–100 mg)	26	1	7%

*Multiple aspartame-containing products used by many individuals.

Table 8-1. *Estimated Consumption of Aspartame-Containing Products by 397 Respondents When Adverse Symptoms Began—continued*

Presweetened pudding or gelatin	Serving (100 + mg)	61	1	15%
Chewing gum	Stick (10 + mg)	47	4	12%

A 62-year-old woman tried one serving of an aspartame soft drink that had been "sent free through the mail." She promptly experienced severe dizziness, tremors, insomnia, and chills "to the point that I wrapped in an electric blanket turned to maximum." These symptoms stopped within one day—only to recur within hours after retesting herself on *four* occasions with either this product or a diet cola. She recalled:

"I could not function fully due to lack of sleep, and the chills made it impossible for me to write or dress normally . . . As a graduate home economist, I recognized that I was experiencing a reaction to something—so started checking out the possible sources."

SEVERITY OF REACTIONS

Statements by the 397 aspartame reactors who completed the questionnaire attest to the severity of adverse effects. *One-half refused to subject themselves even to a single rechallenge for fear of irreparable damage to their health.*

Data from the FDA's Adverse Reaction Monitoring System (ARMS) underscore the potential severity of such reactions. Eight percent of 3,326 complainants were classified as Type I (Tollefson 1987). It was defined as including, but not limited to, "severe respiratory distress or chest pain; cardiac arrhythmias; anaphylactic or hypotensive episodes; severe gastrointestinal distress such as protracted vomiting or diarrhea leading to dehydration; severe neurological distress such as extreme dizziness, fainting, or seizures; or any reaction requiring emergency medical treatment."

Some patients specifically attributed severe reactions to the use of a more concentrated form of an aspartame product. Several made pointed reference to the onset or dramatic exacerbation of such complaints after consuming "100 percent" aspartame.

A 31-year-old secretary drank up to four cans of a diet cola daily. She wrote, "When aspartame was switched from the blend to full strength, I passed out after dinner three days later."

Ordeal By Rechallenge

A surprising number of aspartame reactors retested themselves *at least ten times* before being convinced. Several kept challenging themselves on more than a score of occasions "just to be certain."

Concomitant Hypothyroidism

I have been impressed by the severity of aspartame reactions in patients with underactive thyroid function (hypothyroidism). It is known that the effects of drugs and chemicals tend to be exaggerated and more prolonged among hypothyroid individuals. This condition was known to exist in 22 cases, but had gone unrecognized in other patients prior to consulting me.

A 20-year-old college student suffered severe intellectual deterioration, sleepiness, intense headaches and aggravated reactive hypoglycemia after using aspartame products. Several of her maternal relatives were known to have a goiter. Previously-undiagnosed hypothyroidism was documented by the finding of markedly elevated thyroid stimulating hormone (TSH) levels. Her response to aspartame abstinence and thyroid (thyroxine) replacement was gratifying.

INADVERTENT RECHALLENGE

Many reactors suffered repeat attacks after inadvertently consuming products they did not realize contained aspartame. A few examples are cited.

A 59-year-old professional experienced recurrent convulsions on three separate occasions when he unknowingly drank aspartame-sweetened beverages.

A 30-year-old computer programmer would promptly develop severe lightheadedness, dizziness and abdominal pain after unknowingly drinking any beverage containing aspartame.

A 51-year-old registered nurse previously had suffered severe itching and a rash after drinking aspartame-containing beverages. She stated:

"I thought about a retest even though I was convinced, but decided against subjecting myself to this discomfort . . . I went to a pool party a year later. It was hot and I sat in a lounge chair at the pool from 3–5 pm with three glasses of diet soda—never gave it a thought. By 10:30 pm, the raised itchy red rash was again present on my chest. I stopped the soda, and used Benadryl and hydrocortisone cream. It left."

A 72-year-old female with known aspartame reactions continued to suffer unexplained headaches, visual problems, and "burning and swelling" of the lips and tongue for months. Although she initially denied the further use of aspartame, it *was* present in the ginger-ale she had been drinking.

A 28-year-old woman "immediately" developed severe abdominal pain after drinking a diet cola. It gradually subsided . . . but then recurred when she tried another brand. Her next attack occurred as she drank a beverage not suspected of containing aspartame.

"A year after I'd stopped using it, we ordered out lunch at work. I had asked someone to get me a diet cola which did not have aspartame at that time. They told me that's what they got me.

(It was in a fountain cup, so I couldn't tell.) It never even crossed my mind that it could contain aspartame until the all-too-familiar cramps returned. So I called the food place and then found out I had an aspartame cola. You can't ask for a better test than that."

It is important to consider less obvious sources of aspartame whenever the question of rechallenge arises. For example, "health products" such as vitamins and laxatives may contain aspartame. One notable offender in this category is a popular "sugar-free, effervescent, natural-fiber laxative" sweetened with aspartame. Paradoxically, some patients using it for gastrointestinal problems—whether constipation or diarrhea—then developed more intense abdominal pain, nausea and even bloody stools. These complaints promptly subsided after shifting to the aspartame-free form.

Once aware of the aspartame-is-everywhere phenomenon, many aspartame reactors become leery of dining out for fear of suffering recurrence of this "restaurant syndrome".

A 57-year-old woman experienced extreme nausea, vomiting, dizziness, violent headaches, mental confusion and abdominal pain within four hours after ingesting any product containing aspartame. She studiously attempted to avoid such products when eating away from home. Even so, she suffered several attacks after having been wrongly reassured that there was no aspartame in the food or beverage served her. As a result, she resorted to carrying plastic bags in her purse and car whenever she travels.

Attitudes Concerning Rechallenge

As noted, most patients with severe reactions to aspartame products asserted that they would "never" knowingly re-expose themselves to this chemical. A few thought they might do so under specific conditions, such as a research project under close observation.

A 60-year-old female vowed that she would "never knowingly take it again!" . . . "I would only consume aspartame under a strictly-supervised medical environment, and even then I would have to think long and hard before giving permission for same."

REACTIONS TO SMALL AMOUNTS

Many patients and correspondents emphasized that drinking, eating or chewing even small amounts of products containing aspartame would promptly and predictably induce a severe reaction. Some children experienced headaches, convulsions, or both, within minutes after chewing gum or an over-the-counter analgesic containing aspartame.

> A 61-year-old businessman suffered severe visual symptoms, confusion and amnesia while playing golf. He had taken one cup of coffee sweetened with an aspartame tabletop sweetener at a fast food restaurant enroute to the course. (It did not stock the saccharin-containing sweetener he had requested.) He recalled milder reactions on two occasions after being served similarly sweetened coffee at a neighbor's home.

> A 19-year-old woman had convulsions that were finally attributed to aspartame. She remained seizure-free for 11 months by avoiding such products. She then was handed a piece of "sugar-free" gum at a ball game. Multiple grand mal convulsions recurred within minutes after chewing it.

> A 45-year-old salesman noted extreme drowsiness after chewing gum containing aspartame. He wrote:

> "Just recently, I discovered as I'm driving my automobile that chewing aspartame gum caused drowsiness after only chewing one-half a stick. It caused me to yawn, and to feel sleepy and weak. Sometimes I have to stop driving and close my eyes for a few minutes."

There are several intriguing explanations for the rapidity of reactions to small amounts of aspartame. First, the reactions may represent an "allergy" to aspartame, its components, or breakdown products (Chapter 5). In his classic analysis of "addictive eating and drinking," Dr. Theron Randolph (1956) emphasized that " . . . small quantities of a specific excitant may effectively perpetuate an addiction response in view of the extreme degrees of specific sensitivity commonly involved."

Another possibility may entail prompt absorption of aspartame from the mouth. It then enters the general circulation (comparable to placing a nitroglycerin tablet under the tongue for the immediate relief of angina pectoris), or even might go directly to the brain. It is known that small molecules can diffuse from the back of the mouth into the brain (Maller 1967).

ASPARTAME WITHDRAWAL SYMPTOMS

Some aspartame reactors who had consumed large amounts of products containing this sweetener were impressed by "withdrawal" symptoms that began several days or weeks after stopping it. Such complaints included severe irritability, tension and sweating. They were generally reduced by eating or drinking sugar or resuming aspartame . . . suggestive of some form of quasi-addiction.

One contributory element could be the caffeine-withdrawal syndrome (Roberts 1971b). Enormous amounts of caffeine often were ingested with the compulsive drinking of diet colas.

This phenomenon might be related in part to increased endorphins (opium-like substances within the brain) that can provoke the excessive consumption of sugar, fat and calories (Atkinson 1987). A similar response has been induced with saccharin (Lieblich 1983).

ASPARTAME REACTIONS IN FAMILIES

The discovery of severe aspartame reactions in close relatives—up to seven—was unexpected, both by myself and my patients or correspondents. *One out of five* aspartame reactors reported such a familial occurrence. Some persons first learned of this phenomenon at family reunions.

I was impressed by the apprehension of grandparents who had suffered reactions to aspartame products, and then learned that relatives may be more prone to such reactions. They understandably expressed concern about the welfare of their grandchildren. A registered nurse poignantly pleaded with her daughter: "Please see that none of my grandchildren are given anything containing aspartame."

Representative Case Reports

A 62-year-old woman suffered a marked decrease of vision in both eyes, insomnia, slurred speech, intense tremors, depression with suicidal thoughts, anxiety attacks, unexplained facial and chest pains, diarrhea, marked joint pains, and severe itching while consuming aspartame. These complaints (except for the impaired vision) dramatically improved after stopping such products.

Her younger sister experienced severe headaches, depression, unexplained chest and joint pains, and palpitations when she drank five cans of an aspartame cola beverage. Her symptoms abated within several days after avoiding it.

Other relatives with severe aspartame reactions included two daughters, a ten-year-old grandson (who became "very hyper" after drinking aspartame), and a brother whose "large sores on the face" promptly healed after abstinence from aspartame.

Her husband also had unexplained heart symptoms and nervousness that subsided after avoiding aspartame.

★ ★ ★

A 59-year-old engineer concluded that aspartame products had been inducing or aggravating several problems. They included recurrent lightheadedness, tremors, marked memory loss, depression, respiratory allergies, intermittent loss of hearing, intense thirst, and recurrent bronchial-sinus infections.

His 23-year-old daughter suffered severe headache, dizziness, lightheadedness, unsteadiness of the legs, depression, "anxiety attacks," and marked frequency of urination the day after drinking up to four diet colas. There was concomitant impaired vision and pain in both eyes. (None of these symptoms occurred while taking saccharin.) She gave a history of migraine. These products provoked her complaints on each of four challenges. She wrote:

"I find that when I have aspartame, usually in soft drinks, I feel unusual: lightheaded, headache, queasy, funny, nauseous. My vision also becomes somewhat distorted, and my balance goes. It's kind of a tunnel vision feeling."

★ ★ ★

A 56-year-old woman complained of markedly decreased vision, ringing in both ears, sensitivity to noise, headache, extreme irritability, and unexplained chest pain while drinking three cans of a diet cola daily. Concerning her sisters, she wrote:

"I have sent two copies of this form to my two sisters living in Ohio, as they both had a much stronger reaction to aspartame than I've had—one with pain and depression, and the other with severe depression.

Their problems disappeared in a few short days after they quit drinking aspartame drinks."

Bilateral Inheritance

The children of families wherein *both* the father and mother evidenced reactions to aspartame products seemed even more vulnerable. None had known phenylketonuria (PKU) (Chapter 29) or a family history of this disorder. A young mother stated:

"My four-year-old son has always reacted to aspartame with personality changes, headaches and sleep. My husband complains of memory loss with just one bottle of pop containing aspartame. I, myself, react with swelling and headaches. I would like to know more about aspartame and why it affects us like this, since we don't have the PKU problem."

Comparable Reactions Among Relatives

Another interesting finding among such relatives was the similarity of their reactions to aspartame products.

Three close relatives developed diarrhea after drinking aspartame soft drinks.

Two 40-year-old identical twin sisters suffered severe abdominal pain when they ingested aspartame.

Two women with aspartame-associated convulsions had consumed aspartame during pregnancy and while breast-feeding. The several children of each mother also developed convulsions.

A 62-year-old woman experienced "immediate difficulty in swallowing" after ingesting an aspartame soda. She wrote, "My throat became paralyzed and I could not swallow. My daughter asked if I had checked for aspartame. When I did, that's when I realized that I was using it." The reason for this inquiry: her daughter also suffered "throat paralysis" from aspartame beverages.

The Notion of Suggestibility

It could be argued that the element of suggestibility must be considered when such "victims" inquired about comparable symptoms or other reactions to aspartame-containing products among close relatives. I agree. (A similar argument focused on correspondents who "discovered" more than one complaint while completing the questionnaire.) On the other hand, the detailed descriptions of reactions by most persons, described in this and other sections, are sufficiently convincing to negate such a sweeping assertion.

The Genetic Link

The nature of this possible genetic link remains to be clarified. One of several inherited characteristics affecting enzymic or other metabolic activities may be operative.

- Ninomiya and associates (1987) suggested that the site of action of the *dpa* gene among certain strains of mice that prefer D-phenylalanine resides in the taste cell membrane.
- There may be one or more deficiencies or genetic variants of an enzyme required to break down the combined amino acids (aspartylphenylalanine) in the intestine.

CHAPTER NINE

Convulsions (Epilepsy)

*The study of causes of things must be preceded by
the study of things caused.*
— DR. HUGHLINGS JACKSON

CONVULSIONS ARE AMONG THE most serious reactions
attributable to aspartame products. There are various classifications
of convulsions—also referred to as epilepsy, seizures and "fits." In
this series of 551 persons with adverse reactions to aspartame products, 80 (14.5 percent) suffered typical generalized (grand mal) convulsions, and 18 (3.3 percent) experienced so-called temporal lobe
seizures.

In each instance, aspartame products were considered either a
causative or aggravating factor based on the following facts:

- The majority of patients developed their first convulsion after
  consuming one or more aspartame products.
- Patients with previously controlled epilepsy unexpectedly
  developed recurrent attacks despite maintenance treatment
  with phenytoin (Dilantin®) and/or other drugs.

A 31-year-old woman with a seizure disorder was controlled
on phenytoin and carbamazepine (Tegretol®) for more than
fiveyears. She developed dizziness, drowsiness, severe depression and three convulsions after consuming four cans of a
diet cola and three packets of an aspartame tabletop sweetener
daily. Her complaints disappeared within two months after

discontinuing aspartame. She remained seizure-free there-after.

- No other obvious underlying cause could be found in most of these patients even though the majority had neurologic consultations and underwent extensive studies. The latter included one or more computerized tomography (CT) scans and magnetic resonance imaging (MRI) studies of the brain, repeated electroencephalograms (EEGs), and even angio-grams of the blood vessels of the brain.
- Most had additional aspartame-associated complaints that also intensified prior to the onset of convulsions. Migraine and related headaches were the most impressive. In fact, *half* of the aspartame reactors with grand mal convulsions who completed the survey questionnaire had suffered prior mi-graine or other severe headaches.
- The seizures disappeared or dramatically decreased after ab-stinence from aspartame products, with or without anti-epi-leptic medication.
- The unknowing consumption of aspartame, whether by in-gestion or the chewing of gum, predictably triggered subse-quent grand mal seizures. The amount of aspartame ingested in some patients was remarkably small. This is illustrated by (1) an infant who developed convulsions when his nursing mother drank an aspartame soft drink, and (2) a young woman believed to have aspartame-related epilepsy who con-vulsed within minutes after chewing *one* piece of "sugar-free" gum.

A 52-year-old bank executive, in prior good health, suffered a convulsion after drinking six cups of an aspartame hot cocoa mix daily on eight consecutive nights. She also became tem-porarily blind. These symptoms disappeared within two weeks after avoiding aspartame. A friend later handed her a stick of aspartame gum in a darkened movie house. She then "fell flat on my face in the lobby."

PERTINENT CLINICAL OBSERVATIONS

Young women with migraine, severe reactive hypoglycemia ("low blood sugar attacks"), and fluid retention (recurrent edema) seem more vulnerable to convulsions when they consume aspartame products.

A tendency to grand mal attacks during sleep after ingesting aspartame before bedtime was noted repeatedly. Explanations for this sequence include delayed allergic reactions and aggravated nocturnal hypoglycemia, as discussed in subsequent chapters.

The finding of a normal electroencephalogram (EEG) in children with unexplained seizures should raise the possibility of aspartame-associated convulsions. This possibility is important because (1) the prognosis tends to be good, (2) medication generally can be limited to one drug, and (3) epileptic drugs sometimes can be discontinued within one or two years.

Alcohol intake prior to the seizures was uncommon. Two women (both nurses) developed grand mal seizures after drinking small amounts of alcohol and considerable aspartame-containing beverages.

Aspartame products may render young children more vulnerable to seizures. For example, a two-year-old with fever suffered seizures within ten minutes after chewing aspartame-sweetened acetaminophen (a commonly used substitute for aspirin). This consideration may be significant to health-conscious mothers who elect to give their infants health products containing aspartame rather than sugar (such as vitamins) in an effort to prevent tooth decay.

Most neurologists and pediatricians have been reluctant to consider the possibility of aspartame-associated seizures. However, it is noteworthly that no precipitating cause of *status epilepticus* (a severe convulsion lasting more than 20 minutes, or recurrent attacks lasting 30 minutes during which the patient fails to regain consciousness) could be found in one-third of previously normal children according to several large recent studies.

Family histories of aspartame reactors (Chapter 8) reinforce an association between aspartame-associated severe headache and convulsions. The occurrence of seizures in each of two such women and their children was mentioned earlier.

A 48-year-old-woman with mild migraine experienced severe headaches after drinking diet beverages. They subsided when she avoided aspartame products. Her 15-year-old daughter suffered two seizures under similar conditions.

Aspartame-containing products may interact with phenytoin (Dilantin®) and other anti-epileptic drugs . . . culminating as either recurrent seizures or phenytoin toxicity.

A 29-year-old male drank 18 (!) 12-ounce cans of diet cola daily for one year before suffering two grand mal seizures. (He broke both ankles during one attack.) Other aspartame-associated complaints included severe headache, slurred speech, and unsteadiness of the legs. When the dose of phenytoin was increased due to persistent symptoms while still ingesting aspartame, he developed phenytoin toxicity—that is, his blood level rose above the safe therapeutic range.

REPRESENTATIVE PATIENT REPORTS

The following aspartame reactors with seizures were personally attended, and illustrate many of the problems encountered.

A 31-year-old registered nurse had enjoyed good health. She was subject to mild fluid retention before her periods. She smoked moderately, but had not taken any drug associated with abuse.

The patient drank one or two glasses daily of an aspartame-sweetened lemon drink *for first time* over a one-week period. She then consumed 67 ounces (two liters) of the same brand's orange drink the following Saturday.

Her husband found her unconscious and convulsing at 4 AM on Sunday. She displayed combative behavior in the emergency room, but later had amnesia for this entire period. Itching and a rash over the neck and chest also developed.

A thorough medical and neurologic evaluation failed to reveal a definite cause for the seizure. Her studies included a CT scan of the brain, an EEG, a drug screen, and measurement of vitamin B12 and folate blood levels. The diagnosis of "idiopathic epilepsy" was made, and phenytoin (Dilantin®) prescribed.

Surprised and puzzled by the episode, this nurse began researching a possible connection between aspartame and seizures in the hospital's medical library. She located two comparable case reports and several references. She also found that "diet" orange beverages tend to contain higher concentrations of aspartame—that is, up to 930 mg per liter (335 mg in 12 fluid ounces.) Her neurologist refused to delve into the matter. Instead, he suggested: "Stop reading!"

Due to concern over significance of the seizure, its impact on her career, and driver insurability, she saw me in consultation. After a detailed review of the case, I concluded that her ingestion of 1,100 mg of aspartame within a ten-hour period— about 18.3 mg/kg body weight—could have triggered the seizure. She was carefully weaned off phenytoin, and had no seizures over the ensuing eight months.

The patient later drank "only three sips" of a beverage believed to be "regular" soda, but actually containing aspartame. She became "very incoherent" the following morning.

Two additional conditions subsequently were uncovered: reactive hypoglycemia ("low blood sugar attacks"), and an apparent recent sensitivity to monosodium glutamate (MSG).

* * *

Tammy (see Chapter 1), a 16-year-old girl, was seen in consultation because of recurrent seizures that had failed to respond to conventional therapy by two neurologists. I found a mitral systolic murmur and click (consistent with mitral valve prolapse), episodic tachycardia (rapid heart action) by 24-hour Holter monitoring, and reactive hypoglycemia two hours after she drank glucose.

The patient was placed on an appropriate diet for hypoglycemia. She progressively improved over the ensuing month.

The seizures then recurred on three occasions—either during the late afternoon, or three hours after her evening meal. The train of events consisted of dizziness, weakness, shortness of breath, violent muscular spasms with altered consciousness, and subsequent extreme exhaustion. Her parents were unable to detect any change in pulse during the attacks. Despite careful questioning, I could not uncover any intercurrent emotional problem or other convincing contributory factors that might have produced these symptoms.

As I examined Tammy in my office at 5 PM, she began to exhibit generalized muscular contractions, facial grimacing and altered consciousness. Her glucose concentration was found to be normal (91 mg per 100 ml). Furthermore, she had snacked two hours earlier.

Her mother volunteered two crucial clues at that point. First, the paternal grandmother had "a severe allergy to aspartame." Second, the patient was drinking considerably more aspartame-containing sodas because of my previous advice to avoid sugar (which can trigger attacks of reactive hypoglycemia). Her snack that afternoon consisted of an aspartame-containing pudding. After this episode, she avoided aspartame products . . . and remained seizure-free for several weeks.

With formal consent, I then rechallenged Tammy with a *small* serving of the *same* pudding containing aspartame. Her blood glucose concentration fell more rapidly at the second hour than during the prior glucose tolerance test. Just before the third hour, she began evidencing confusion, marked muscular jerking, and a severe sweat. The test was terminated at that point.

No seizures developed over the ensuing eleven months of abstinence from aspartame products.

★ ★ ★

A 29-year-old businessman sought consultation because of recurrent grand mal seizures over an 18-month period. He had begun drinking considerable amounts of diet soft drinks and eating other aspartame products six months before the first convulsion. He suffered five major attacks even while on relatively large doses of phenytoin and carbamazepine. Other complaints included frequent bowel movements, severe fatigue (even after a full night's sleep), pathologic drowsiness, and a 30-pound weight gain. He neither smoked nor drank alcohol.

When first seen by me, this patient already had *five* CT scans of the brain, *three* MRI studies, and *five* electroencephalograms. The brain studies presumably indicated "inflammation of either the white or gray matter of the brain." He was hospitalized several times at different medical centers.

Physical examination revealed only slight elevation of the blood pressure in a modestly overweight male. Additional studies were normal except for persistent elevation of the blood glucose concentration at 3 hours and 3-1/2 hours during a glucose tolerance test. (He still was on phenytoin, which could have influenced these results.)

The patient had no further seizures for six months after stopping *all* aspartame products. During this time, his blood pressure remained normal.

In discussing the case with his neurologist, I suggested a gradual reduction in the dosage of the two anti-epileptic drugs, especially in light of the patient's fear of potential side effects. The physician not only adamantly refused to do so, but told the patient that aspartame "has absolutely nothing to do with your convulsions."

He remained seizure-free and otherwise asymptomatic after avoiding aspartame products. A repeat MRI study—performed at my hospital to clarify the abnormality noted in other institutions— was interpreted as consistent with a *congenital* vascular malformation of the brain.

OTHER EVIDENCE

FDA Registry

As of early 1987, 137 consumers had *volunteered* reports to the FDA in which they attributed their convulsions to aspartame products (Tollefson 1987).

Dr. David G. Hattan (Chief, Regulatory Affairs Staff at the FDA Office of Nutrition and Food Sciences) subsequently provided me with a summary of the information up to July 1987 on 149 consumers with alleged aspartame-induced convulsions. The data appear in Tables 9-1 and 9-2. It is noteworthy that the majority (87.2%) of reactions were classed as Type I (most severe category). The majority of these complainants also cited other aspartame-associated problems.

> The classifications used by the FDA for determining feasibility causation of incriminated products containing aspartame are as follows: *Types* (according to severity of the seizure) I—severe II—moderate; *Groups* A—recurrence of seizures each time different products containing aspartame were consumed B—recurrence each time the same product was consumed C—symptoms associated with the ingestion of such products, but without rechallenge by the complainant D—failure of symptoms to recur every time these products were consumed, or if a consultant-physician regarded the association as unlikely.

Table 9-1. *Convulsions Allegedly Associated with Aspartame Products (General Data on 149 Cases in FDA Registry)*

Sex		
Females	96	(64.4%)
Males	53	(35.6%)
Race		
White	72	(48.3%)
Nonwhite	6	(4.0%)

Table 9-1. *Convulsions Allegedly Associated with Aspartame Products*
(General Data on 149 Cases in FDA Registry)—continued

Unknown	71	(47.7%)
Age Groups (years)		
1–19	32	(21.5%)
20–29	28	(18.8%)
30–39	42	(28.2%)
40–49	21	(14.1%)
50–59	5	(3.4%)
60 +	6	(4.0%)
Unknown	15	(10.1%)
Time From Ingestion of Aspartame		
"Immediate"	24	(16.1%)
1–12 hours	68	(45.6%)
13–24 hours	8	(5.4%)
24 hours	49	(32.9%)
Type of Reaction		
I	129	(87.2%)
II	19	(12.8%)
Unspecified	1	
Group		
A	25	(17.0%)
B	21	(14.3%)
C	29	(19.7%)
D	72	(49.0%)
Unspecified	2	
Aspartame Products Consumed		
Soft Drinks	115	(77.2%)
Tabletop Sweeteners	39	(26.2%)
Hot Chocolate	13	(8.7%)
Iced Tea	12	(8.1%)
Gum	11	(7.4%)
Soft Drink Mixes	40	(26.8%)
Puddings/Gelatins	16	(10.7%)

Table 9-2. *Alleged Aspartame-Associated Convulsions (Concomitant Reactions in 149 Cases from FDA Registry)*

Headaches	28	(18.8%)
Depression/Mood Changes	23	(15.4%)
Other Neurologic Features	18	(12.1%)
Dizziness	14	(9.4%)
Abdominal Pain/GI Complaints	12	(8.1%)
Memory Loss	9	(6.0%)
Weakness/Fatigue	7	(4.7%)
Visual	6	(4.0%)
Fainting	6	(4.0%)
Hives/Rashes	5	(3.4%)
Sleep Problems	5	(3.4%)
Heart	3	(2.0%)
Bone/Joint Pain	1	(0.7%)
Menstrual	1	(0.7%)

Reports By Other Physicians

Dr. Richard Wurtman (1986), a researcher at the Massachusetts Institute of Technology, stated that he had been contacted by more than 100 persons with alleged aspartame-associated seizures. He also was impressed by the frequency of previous migraine in such individuals, and the intensification of their headaches prior to convulsions. Furthermore, his experimental studies indicate that low doses of aspartame enhance seizures in animals predisposed to abnormal brain activity (Wurtman 1987c).

Another physician informed me about a commercial pilot who had lost his license because of unexplained convulsions. Deducing they probably were triggered by aspartame beverages, he abstained from such products . . . and became seizure-free. In an attempt to document such *specific* intolerance and regain his pilot's license, he *purposefully* rechallenged himself with an aspartame soft drink. Another seizure promptly ensued.

Consumers' Reports

My appearance on radio and television shows predictably evoked letters from persons who had attributed their seizures to aspartame products. The following information, derived from this correspondence, is of course wholly anecdotal and could readily be challenged. When viewed as part of the larger picture, however, the pattern is remarkably consistent with the other evidence.

This letter was received after I had been interviewed on Philadelphia's radio station WWDB.

"On November 5, 1985, I suffered a seizure at home. I was rushed to the hospital and stayed there for four days. I had blood work, CT scans and EEG. I was discharged and told I had an epileptic condition and low magnesium. I have been on 300 mg of Dilantin® ever since. I also had my driver's license recalled for one year, and as of today it has not been reinstated.

"I was very upset with the epilepsy label that was laid on me. It was very hard to accept . . .

"I consumed large amounts of aspartame as soda, coffee, and tea. The day I had my seizure my intake of aspartame was quite a lot. I am totally convinced that it triggered my seizure. I have not had a seizure since not using aspartame in any form."

★ ★ ★

A 43-year-old executive suffered a grand mal seizure in January 1986. Extensive medical and neurologic evaluations proved inconclusive. He sent me the following letter after being told about one of my television interviews.

"All test results were normal. Since I am 43 years of age and of good health, with no family history of seizures, the cause of my problem was inconclusive. One of the mysteries of life so to speak.

"I did point out (to the physicians) that I had begun a diet some three weeks before the seizure. I believed it to be a sensible one, whereby I simply reduced my caloric intake—eating the same foods but less. What I did not point out (at the time there was no reason to be suspect) is the fact that I had substituted for my normal beverage of milk an array of diet drinks. Namely, they were two diet colas, and aspartame-sweetened coffee and tea. While I don't normally consume diet beverages because I find the taste repugnant, I did drink them regularly during the three weeks prior to my seizure. As a matter of fact, I continued to use these products for quite some time later.

"It may or may not be relevant here, but I also continued to experience auras despite the fact that I was taking 500 mgs of Dilantin®, and my blood Dilantin® count was 14."

★ ★ ★

A 59-year-old forester had his initial seizure during March 1984. He was consuming three packets of an aspartame tabletop sweetener and up to eight glasses of an aspartame soft drink daily. He then suffered two convulsions in July 1984 after unknowingly ingesting aspartame. Severe convulsions recurred in March 1985— again after taking aspartame in an unrecognized form.

His wife described their ordeal when he went on a business trip during July 1985 to evaluate timber.

"It was a hot day and a tiring trip, so he bought two cans of a soft drink, which he did not realize had changed to aspartame from saccharin that week. He drank them, and later on the way home stopped and drank two glasses of punch which we later found out was made from aspartame. When he got home at 7 p.m., he wouldn't eat any dinner and was cranky and tired, and went right to bed. At about 5:30 a.m., I was awakened by this roaring sound. He was having another violent convulsion . . . he dislocated his shoulder and was in lots of physical pain. I asked the doctor if this didn't prove that it was the aspartame causing his convulsions. He said "Maybe," I said, "Shouldn't someone write the aspartame company and tell them what had happened?" All he said was, "Yeah, you do it." We asked to see a neurologist. He was even less interested in aspartame . . .

"A week later, we went out to dinner at a local restaurant with friends. While we had a cocktail from the bar, he ordered a regular soft drink. They evidently used the nozzle with the brand's aspartame drink . . . We came home and went to bed. About 2:30 a.m., he went into the first of seven consecutive convulsions."

SOCIOECONOMIC IMPACT

The economic, employment and social ramifications of aspartame-associated convulsions can be enormous, as might be inferred from the case reports.

Accidents and Licensure

Since seizures, drowsiness and "blackouts" contribute to driving accidents (Roberts 1971b), they become major considerations in obtaining or retaining driver licensure and insurance.

- Many drivers and several pilots with aspartame-associated seizures have lost their licenses. An Air Force pilot told a Senate hearing on November 3, 1987 that he was permanently grounded because of a seizure which developed after consuming one gallon of an aspartame beverage daily (Collings 1988). There had been no recurrence of seizures over several years of abstaining from aspartame products.
- The fatal automobile accident of a 21-year-old male apparently resulted from a seizure and associated loss of consciousness while driving. He was consuming up to four gallons of diet cola daily, and had many suggestive aspartame-associated complaints. His sister, who relayed this information, also was an aspartame reactor.

Consumer Frustration With Physicians,
the FDA, and Epilepsy Organizations

Many patients with aspartame-associated convulsions were frustrated by physicians who denied, ignored or trivialized the possibility of seizures due to aspartame products. The following issues were especially annoying.

- The implication that "the aspartame thing is primarily in my head, and perhaps I should see a shrink."
- The insistence by neurologists that high doses of anti-epileptic drugs be continued indefinitely even though the patient had remained seizure-free after avoiding aspartame products.
- The reluctance of neurologists and epilepsy-oriented organizations to provide information and research findings that patients had requested about these and other aspartame-associated reactions.

Such frustration was compounded by the denial of any link between seizures and aspartame products from the FDA and epilepsy organizations.

- The Chairman of the Professional Advisory Board of *The*

Epilepsy Institute distributed the following letter on March 18, 1986.

"Recent publicity on aspartame and seizures may have promoted calls from concerned patients. As an organization devoted to people suffering from seizure-related problems, we at The Epilepsy Institute investigated this allegation and found aspartame to be safe for people with epilepsy . . . as Chairman of The Epilepsy Institute's Professional Advisory Board, I have also reviewed the data and see no cause to link aspartame and seizures."

• The FDA offered this response to my August 1986 letter.

"We have reviewed a total of 141 reports of seizures and have found no association between occurrence of seizures and exposure to aspartame-containing products . . . In addition, clinical studies are being supported by the sponsor of aspartame to assess, under appropriately controlled conditions, whether it causes or may contribute to the occurrence of seizures in individuals who believe they are sensitive to aspartame. At this time, no evidence has been established for a consistent seizure pattern or dose response relationship."

• Dr. Harold E. Booker, Chairman of the Professional Advisory Board of the Epilepsy Foundation of America, was requested by the United States Senate Committee on Labor and Human Resources to make a statement concerning any possible risk to epileptic patients from aspartame. His correspondence, dated October 31, 1987, stated: "As of the last meeting of the Board (April, 1987), their report was that, *in their opinion,* there was no valid, scientific evidence that aspartame posed any risk for people with epilepsy." (Italics supplied)

MECHANISMS BY WHICH ASPARTAME MAY CAUSE CONVULSIONS

Some of the mechanisms by which aspartame may cause or aggravate convulsions have been mentioned in Chapters 5 and 6. They include high phenylalanine and aspartic acid concentrations (including their stereoisomers) in the brain, the alteration of neurotransmitters (especially serotonin and norepinephrine), the effects of methyl alcohol, and exaggerated hypoglycemia ("tissue glucopenia"). Aspartic acid and its derivatives or analogs are powerful convulsants when injected into the brain (Turski 1984).

Other aspartame-related influences, operating in tandem, might serve to increase the potential for seizures.

- Chronic sleep deprivation due to aspartame-associated insomnia depletes norepinephrine—a phenomenon known to provoke seizures in animal models.
- The consumption of excessive caffeine (as with aspartame-sweetened coffee or diet colas) can lower the threshold to convulsions, as indicated by experimental pentylenetetrazol-induced seizures (Maher 1987).
- The intake of excessive fluid and sodium may contribute to seizures. Aspartame-induced intense thirst (Chapter 21), especially during hot weather, is noteworthy in this context.

RESEARCH: THE BOTTOM LINE

The ultimate "experimental model" for studying the problem of aspartame-associated seizures is the vulnerable brain of a *human* aspartame reactor. Stated differently, *no single animal model can accurately convey the many facets of epilepsy in man.* This means that performing studies on monkeys and other animals in an attempt to answer the question definitely will not succeed. The results cannot be accurately extrapolated to humans.

As stressed in Chapters 4 and 5, the documentation of convulsions attributed to aspartame-containing products, and the scientific investigation of such patients, *must* be done using the *same* products obtained from conventional retailers, or as prepared for customary use. The continued administration of pure aspartame capsules and freshly-prepared cool beverages for double-blind studies, as has been done, is likely to add to the existing confusion.

CHAPTER TEN

Headache

Nothing is obvious to the uninformed.
— ANONYMOUS

HEADACHES ARE THE MOST common symptom attributed to aspartame products. Almost *one out of two* persons in this series—specifically, 45 per cent—experienced severe headache (Roberts 1987a,b; Roberts 1988e). While it could be argued that headache is a common complaint regardless of the aspartame issue, its incidence in the general public is far lower . . . estimated at only ten per cent.

Most aspartame reactors who developed headache were impressed by (1) the prompt improvement following abstinence from aspartame-containing products, and (2) the dramatic recurrence of headache, often within minutes or several hours, when these products were resumed. (In fact, it was an unexpected headache that repeatedly alerted some persons to the likelihood they had inadvertently ingested aspartame.)

A striking three-to-one *female preponderance* of headache sufferers was found.

Many individuals who developed headaches had been subject in the past to migraine and its variants, including so-called cluster headaches. Some emphasized, however, that their aspartame-associated headaches were "different."

Another subgroup of reactors stressed that they had not been previously subject to *any* type of headache.

The effects of aspartame and phenylalanine on major neurotransmitters within the brain (Chapter 5), have considerable pertinence to migraine and other "vascular" headaches.

CORROBORATION OF FINDINGS

Admittedly, there are many different causes for headache. It is therefore fair to ask for corroboration of my observations by other data.

FDA records contain many more cases. This agency received 951 volunteered complaints of headache from consumers who alleged that they were caused or aggravated by aspartame products (Tollefson 1987).

The cause-and-effect relationship between aspartame intake and the precipitation of migraine has been confirmed in *a controlled double-blind randomized cross-over study* by Koehler et al (1987).

Positive responses to provocative tests in the form of induced headache also have been described by other physicians.

- Ferguson (1985) reported a 22-year-old woman who had severe headaches while using about ten packets of an aspartame-containing artificial sweetener. They subsided within several hours after stopping aspartame. This sequence recurred during five separate trials—namely, the headache predictably returned (with concomitant flushing and sweating) when the subject resumed aspartame.
- A 31-year-old female studied by Johns (1986) developed "throbbing vascular headaches with associated gastrointestinal symptoms" within 1 to 2 hours after drinking two or three cans of a diet cola. They also recurred within 1 1/2 hours following rechallenge with pure aspartame (500 mg in 14 fluid ounces), but not with comparable amounts of saccharin or sugar (sucrose).

ASSOCIATED CLINICAL PROBLEMS

Erroneous Diagnoses

Dozens of patients in my series initially were suspected of having some more serious disease due to the severity of their headaches.

The diagnoses included brain tumor, cerebral artery aneurysm, temporomandibular joint (TMJ) dysfunction, cervical disc degeneration, and various types of neuropathy ("neuralgia") involving the cranial and cervical (neck) nerves. Accordingly, nearly all had CT scans or imaging (MRI) studies of the brain—especially when there were concomitant neurologic and psychiatric complaints (e.g., confusion, memory loss, dizziness, visual impairment, personality change.)

Some individuals even were subjected to lumbar puncture and invasive arteriography of the cerebral blood vessels (entailing the injection of dye into a major artery within the neck) in search of an "organic" cause for the headaches. Cerebral angiography in patients with migraine is not innocuous, however, since the procedure has resulted in new focal neurologic complications.

Concomitant or Ensuing Epilepsy

The continued use of moderate or large amounts of aspartame products by patients with recurrent severe migraine frequently preceded a convulsion (Chapter 9). Thirty persons completing my questionnaire (Chapter 7) who experienced headache *and* convulsions while consuming aspartame gave a history of migraine or other severe headaches.

Women subject to considerable premenstrual fluid retention and/ or severe reactive hypoglycemia (see below), and those taking an oral contraceptive drug ("the pill") also appear to be at increased risk for both disorders.

Hypoglycemia

Elsewhere, I have emphasized the importance of reactive hypoglycemia ("low blood sugar attacks") in precipitating migraine and related vascular headaches (Roberts 1964a,b; 1967d; 1971b). The occurrence of these headaches on arising, or even later during the day, commonly comes in the wake of a preceding, albeit unrecognized, hypoglycemic attack during sleep.

Aspartame consumption can aggravate hypoglycemia for several reasons. These include increased release of insulin by its component amino acids, the marked reduction of calories during attempted weight loss, prolonged delays in eating, excess insulin release associated with "the pill," "hepatic hypoglycemia" from concomitant intake of alcohol, and oral hypoglycemic drug therapy for diabetes.

"Chemical Headache"

Chemical headache is a well known clinical entity. Some of the provocative substances are cheese, citrus fruits, alcohol, nitrates, nitrites ("hot dog headache"), monosodium glutamate (MSG), tyramine, and phenylethylamine ("chocolate headache").

Several patients in this series added aspartame to the list of chemicals capable of triggering their attacks of headache—especially when taken in conjunction with MSG or wine. A 48-year-old woman wrote

> "I developed classic migraine (flashes of light, numbness on one side of my body, impairment of my power to use words, nausea, vomiting, etc.) immediately after drinking beverages sweetened with aspartame. I have discovered over the years that the same thing happens when I ingest phenylpropanolamine found in some cold pills and some diet pills. Wine flavored with woodruff also gives me migraine headaches. Obviously, I have learned to completely avoid aspartame, phenylpropanolamine and woodruff!"

Aspirin Use

The use of substantial amounts of aspirin for relieving aspartame-associated headache invited other complaints. Aspirin and various salicylates can irritate the inner-ear (eighth cranial) nerve. They therefore may induce dizziness, ringing of the ears, or hearing loss. Aspirin also can cause severe gastrointestinal irritation and may aggravate reactive hypoglycemia.

Continued Aspartame Use

Further evidence for a quasi-addiction to aspartame products emerged from this aspect of my study. Twenty reactors who completed the questionnaire indicated they *still* consumed them despite the persistence or recurrence of severe headache!

REPRESENTATIVE CASE REPORTS

A 24-year-old woman suffered increasingly severe headaches while consuming up to three liters of an aspartame cola daily. She deduced aspartame to be the cause after viewing a feature on this subject. "If it wasn't for the Consumer's Report Special, I might have never associated my severe headaches (every day) with the aspartame. Only God knows what would be wrong with me now. The night I saw the show, I immediately stopped drinking diet drinks. The next night the headaches were gone."

★ ★ ★

A 27-year-old ophthalmic assistant experienced progressive and severe unilateral headache for two months. There was concomitant dizziness, numbness and nausea. A neurologist diagnosed her in an emergency room as either "having an aneurysm, or your problem may be related to stress." She had been drinking up to eight cans of an aspartame cola and eight one-liter bottles of the same beverage daily.

Reluctant to accept either diagnosis, she pursued the matter in detail.

"From that point on, I started making checklists to see if anything else could be causing the pain. Dentists, sinus and ophthalmic examinations all checked out okay. Then I took a look back two months ago to see if anything had changed in my diet. That's when I realized I had started drinking the new diet cola with aspartame in it . . . That night, I stopped drinking the pop and the pain began to diminish, slowly, day by day. (In one week, I was) not experiencing any dizziness or nausea, and I didn't feel like I'm in a fog anymore . . . It is now three weeks since I ceased the drinking of colas. I am today experiencing almost no symptoms except an occasional burning sensation on my scalp. I know now that it was definitely the aspartame."

CHAPTER ELEVEN

Confusion and Memory Loss; Intellectual Deterioration

You must still be bright and quiet And content with the simple diet . . . But the unkind and the unruly, And the sort who eat unduly, They must never hope for glory—Theirs is quite a different story!
— *R O B E R T   L O U I S   S T E V E N S O N*
(Good and Bad Children)

CONFUSION, MEMORY LOSS, AND even profound intellectual deterioration have been impressive among patients in this series who exhibited adverse reactions to aspartame products. Fortunately, most of these symptoms improved when aspartame products were avoided.

A controversial question necessarily must be raised: "Does this have *any* connection with Alzheimer's disease?" The subject will be discussed in Chapter 28.

CONFUSION AND MEMORY LOSS

More than one out of four aspartame reactors in this series—28.5 per cent—experienced marked confusion, gross impairment of memory, or both. Prompt improvement generally occurred within several days or weeks after abstinence from aspartame products. These complaints *predictably* recurred on rechallenge . . . even with small amounts.

These observations are corroborated by *FDA data* (courtesy, Dr. Linda Tollefson). More than 125 aspartame consumers complained to this agency about experiencing memory loss while using aspartame products.

The mechanisms by which aspartame and its components or breakdown products may affect essential memory circuits within the brain are mentioned in Chapters 5, 6, and 28, and in my reports thereon (Roberts 1988a, 1989b).

Age Range

The age range of the 120 persons who indicated "severe memory loss" in the survey questionnaire was as follows:

> 10–19 years . . . 4
> 20–29 years . . . 14
> 30–39 years . . . 26
> 40–49 years . . . 23
> 50–59 years . . . 18
> 60 and older . . .17

Such unexplained memory loss was understandably frightening. Many active adults expressed concern about possibly having an early stage of Alzheimer's disease (see below). The prompt clearing of confusion and return of memory after aspartame abstinence proved reassuring.

Aspartame-associated confusion was impressive even among some teenagers.

> An 18-year-old male forgot where he was or where to turn, while driving in his own neighborhood when drinking two liters of aspartame-containing soft drinks daily.

Misdiagnoses

Erroneous diagnoses as to the cause of these mental changes were the rule. Most patients with aspartame-associated confusion and memory loss had undergone CT scans and MRI studies of the brain, EEGs, lumbar puncture, and a host of other tests.

Since these symptoms are *nonspecific,* they can occur in scores of other conditions. A frequently-encountered misdiagnosis was the possibility of "side effects" from medications. For example, beta-blockers (such as propranolol or Inderal®), commonly prescribed for hypertension and angina pectoris, were repeatedly—but erroneously—implicated before an aspartame reaction was suspected.

> A 62-year-old businessman had been taking the same dose of propranolol as treatment for angina pectoris over a considerable period. He complained of increasing confusion, memory loss, and visual difficulty while consuming three packets of an aspartame tabletop sweetener daily. His mental clarity normalized within two weeks after stopping this product *without* any change in the medication.

Occupational Impact

Aspartame-associated confusion and memory may have serious professional, business and personal consequences.

- Aspartame reactors with these symptoms often mentioned diminished confidence in continuing, or resuming, their professions and occupations. Registered nurses were particularly apprehensive in this regard.
- Aspartame-related confusion and memory loss interfered with the ability of seasoned instructors to teach at a high level of proficiency.
- Well-known hosts of radio talk shows and persons anchoring television programs admitted to having been gripped by fear over the threat to their careers because of confusion, memory gaps and swings of mood before they stopped drinking diet colas and aspartame-sweetened coffee.

Manifestations

The severity of aspartame-associated confusion and memory loss ranged from "small forgettings" to profound mental disability. Some examples:

- Several aspartame reactors volunteered that they occasionally couldn't remember *their own names*!
- Some referred to the problem as "difficulty in concentrating."
- A 30-year-old consultant reported, "I could not consciously remember what I had done or said before."
- An engineer had difficulty recalling "names and descriptive adjectives."
- A building inspector stated, "I could not think straight."
- A 47-year-old woman was extremely dismayed over the loss of her "photographic memory."
- Experienced secretaries made many typing and computer errors.
- A 34-year-old teacher "began having trouble sequencing assignments" after starting to drink two cans of diet cola daily.

Several patients who predictably experienced confusion and memory impairment after consuming aspartame-containing products promptly related to the cartoon depicted in Figure 11-1.

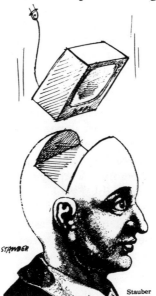

FIGURE 11-1

Reproduced with permission of *World Press Review* (September 1986) and Mr. Stauber of *Nebelspalter* (Zurich)

Representative Case Reports

The 45-year-old superintendent of a large school system began drinking one can of diet cola daily for weight control. He noted predictable "inability to concentrate fully, lightheadedness, and headache feelings" within two hours . . . a sequence reproduced on *eight* rechallenges. Each time, his symptoms disappeared within two days after avoiding this beverage.

★ ★ ★

A 58-year-old man returned before his scheduled annual checkup because of profound memory loss and confusion. These symptoms had become "so bad that if I am cooking something in the kitchen, I'm likely to forget about it." He also suffered unexplained headaches, dizziness, decreased vision, and depression over the previous month. Considerable amounts of aspartame-containing beverages were being ingested. His complaints virtually vanished within one week following abstinence from aspartame, and have not recurred during seven months of personal observation.

★ ★ ★

A 57-year-old executive began consuming aspartame cola in 1983. His consumption progressively increased to 10 to 12 cans daily, especially in hot weather. He developed impairment of memory and associated speech difficulties, as well as severe arthritic pains, insomnia and increasing nervousness. He wrote

"I am an executive in a large U.S. company. My job requires me to make frequent presentations and talk on my feet. During this period I was virtually unable to talk, let alone speak intelligently. Worse, I often couldn't remember *any* of the names of my peers. If I didn't have a psychosis before, the combined speech and memory problems created one . . . the loss of memory is a strange mix of long and short term. I cannot remember a single joke, yet I can remember telephone numbers for even rarely-called numbers."

Within two weeks after avoiding aspartame, he felt markedly improved and was able to discontinue various prescribed medications.

★ ★ ★

A 35-year-old female psychotherapist experienced a recurrent problem with memory loss and lightheadedness promptly after drinking an aspartame cola beverage. She described the reaction as a "spacy feeling and hard to concentrate". She retested herself more than four times before being convinced of the relationship. Concomitant side effects included depression, marked diarrhea, severe bloat, itching, and frequency of urination. She commented: "I couldn't think straight. Being a therapist, this is not alright."

INTELLECTUAL DETERIORATION

I have observed severe intellectual deterioration associated with the use of aspartame products. It was usually manifest as great difficulty in reading and writing, obvious problems with memory, and grossly impaired orientation to time, place and person. In some patients, these features cleared slowly after the avoidance of aspartame.

A 29-year-old female public relations executive experienced severe impairment of reading while ingesting three packets of an aspartame tabletop sweetener and one glass of aspartame-sweetened hot chocolate daily. She described "letters transposing themselves." She "thought I was losing my mind." Other complaints included visual problems, severe headache, irritability, palpitations, a rash, and marked thinning of her hair. All improved within one week after avoiding aspartame . . . only to recur within several days following each of two challenges.

The personal, professional, occupational and health ramifications of such intellectual deterioration often were poignant. They became compounded when (1) concomitant medical disorders (especially hypothyroidism and hypoglycemia) remained undiagnosed or untreated, and (2) persons with reading disability and other forms of brain dysfunction (including stroke) were encouraged to use these sugar-free products. One attendant anguished: "I work at a residential facility for multi-handicapped young adults. I am concerned about the amount of aspartame consumed there."

Representative Case Reports

I saw an 18-year-old college student in consultation in December 1986 because of profound intellectual deterioration. It followed her use of aspartame products for "weight control." She had become incapacitated academically—with a drop in I.Q. of 20 points.

Prior to aspartame consumption, she had been an outstanding student at a major university, as well as a skilled typist and pianist. The director of an honors program had written her during the previous semester: "On the basis of your performance (in economics), you appear to be an excellent candidate for the honors program."

Other complaints included severe headache, decreased vision in one eye, dizziness, intense drowsiness, tremors, insomnia, depression with suicidal thoughts, severe itching, burning on urination, a decided change in personality, abdominal pain, recurrent nausea, the cessation of menstrual periods for six months, and a paradoxic gain of 15 pounds.

Many physicians had previously seen the patient. Her studies included two CT scans and one MRI study of the brain, and numerous tests *except* those described below. Multiple medications were prescribed for a presumed primary depression and possible organic brain syndrome, but proved ineffective. Extensive neuropsychologic testing at a major university concluded:

"Significant findings include moderate psychomotor slowing, moderate memory impairments, mild expressive language deficits, and impaired nonverbal reasoning and problem-solving skills with relatively well-preserved verbal processing abilities. These findings are consistent with bilateral cerebral dysfunction in the region of the temporal lobes, maximal in the right cerebral hemisphere. These findings are directly correlated with EEG findings of bilateral temporal sharp activity. The observed pattern of neuropsychological deficits is *not* consistent with that seen in patients with primary psychiatric disorder . . . The patient performance on neuropsychological and personality measures is *not* consistent with a diagnosis of schizophrenia."

I obtained the following additional information.

- She experienced pathologic drowsiness after taking an aspartame drink, as evidenced by falling asleep in class about 9 AM and during the early afternoon. She also had dozed while driving—once being stopped in the morning by a policeman for suspected inebriation.
- She was subject to attacks of severe weakness ("draining of my energy") and intense hunger, especially after exercising.
- Marked "twitching" and tremor of her hands often precluded typing or playing the piano.
- She had developed a markedly dry skin and severe sensitivity to cold.
- Her family history revealed several close relatives with both diabetes and goiter.

On physical examination, I noted dry skin and overt "restless legs." Further studies documented the presence of reactive hypoglycemia and hypothyroidism. Her thyroid stimulating hormone (THS) level was 13.2 (normal 1–5). The glucose tolerance test had to be stopped at 3-1/2 hours because of a severe clinical hypoglycemic reaction.

The patient was advised to avoid aspartame and follow an appropriate anti-hypoglycemia diet. She also was begun on thyroxine (Synthroid®). There was subsequent clinical improvement, including the resumption of normal menses. Her intellectual deterioration persisted, however, requiring her placement in a facility for the mentally retarded.

★ ★ ★

A 23-year-old female student evidenced extraordinary mental deterioration after she began consuming aspartame-containing cola beverages. While in high school, she had won a Telluride Association Scholarship in competition with more than one million students from the entire country. Similarly, she scored in the 98th percentile nationally in all areas on her GMATs. She began an excellent job as a financial analyst for a large corporation in April 1986. In addition to profound confusion and memory loss, she suffered severe drowsiness, petit mal seizures, tremors, tingling and numbness of the extremities, anxiety, personality changes, visual impairment in both eyes, sensitivity to noise, attacks of rapid heart action, abdominal pain with nausea and bloody diarrhea, joint pains, and the loss of menstrual periods.

Her mother also suffered reactions to aspartame products, chiefly as migraine attacks. She described the extent of her daughter's mental and intellectual deterioration in this excerpted correspondence.

"My daughter's reaction to products containing aspartame has been far more serious than mine for several reasons. For nearly two years, she was unaware of the possible connection between her medical problems and aspartame . . . She continued to consume aspartame and her condition worsened.

"She was living in another state, completing her graduate studies, so we did not have the opportunity to see her very often. However, her behavior became more and more bewildering . . . Finally, and nearly too late, she announced (April 1986) that she was experiencing bizarre symptoms with increasing frequency. She said that she thought she was losing her mind, so she had consulted a neurologist (late 1985, early 1986). He diagnosed her as having temporal lobe epilepsy.

"After my daughter told me of her symptoms (April 1986), it suddenly occurred to me that aspartame could have triggered her seizures, etc., since I knew that beverages sweetened with aspartame caused me to experience migraine attacks. I explained my theory to her over the telephone and in letters . . . Only after she stopped consuming products sweetened with aspartame did she realize that aspartame had ruined her physical and mental health."

The mother informed me in a subsequent telephone conversation that her daughter continued to avoid aspartame products, was improving almost daily, and had reached the point of applying for a Ph.D. program.

CHAPTER TWELVE

Other Neurologic Reactions

Dizziness; Shaking (Tremors); Severe Sleepiness (Narcolepsy); Hyperactivity; Atypical Pains (Neuralgia, Neuropathy); Myasthenia; Slurred Speech; Insomnia

The ultimate significance of any observation can, therefore, be evaluated only over a long period of time. The only advice that can be given is that scientists accept the responsibility of reporting and interpreting their findings in whatever way their best judgment indicates.

— *N. J. VIANNA (1975)**

THE WAY IN WHICH the brain and peripheral nerves show their vulnerability to products containing aspartame, and its components or metabolic breakdown products, are not limited to seizures, headache and memory impairment. Others symptoms include dizziness, tremors, sleepiness, hyperactivity, atypical pains, slurred speech, and muscle weakness. These neurologic features, as encountered in 551 aspartame reactors, are reviewed in this chapter.

The frequency with which an *erroneous diagnosis of multiple sclerosis* was made in aspartame reactors deserves special attention. This is particularly true among weight-conscious young women who develop visual and neurological problems while consuming considerable amounts of aspartame products. In my opinion, this diagnosis ought be deferred at least several months after abstinence from aspartame to enable sufficient observation for spontaneous recovery.

DIZZINESS AND UNSTEADINESS

One out of three aspartame reactors in this series—39.4 per cent—complained of dizziness, unsteadiness, or both. The FDA had received 419 complaints of comparable aspartame-related dizziness and instable gait or posture as of early 1987 (Tollefson 1987).

The clinical descriptions ("giddiness," "vertigo") varied, as did the events precipitating these symptoms. The problem commonly occurred while the patient looked down—as when washing one's hair.

Improvement generally occurred within several days or weeks after avoiding aspartame-containing products; occasionally, it was delayed for several months. The use of aspirin for aspartame-associated headache or joint pain aggravated dizziness in some individuals.

The emotional impact of severe and persistent dizziness was often considerable until the contributory role of aspartame products was recognized. Patients with high blood pressure, coronary heart disease and diabetes, for example, worried about possibility of a small stroke, an atypical insulin reaction, or a brain tumor.

Representative Case Reports

A 30-year-old computer programmer suffered recurrent attacks of lightheadedness and abdominal pain while drinking two cans of an aspartame cola daily. He deduced this beverage to be the cause.

"Within minutes of drinking one diet cola, I feel lightheaded. In 15 to 30 minutes, I feel better. I have switched back to a regular cola. Every now and then I'll accidentally buy a diet soft drink, and will be dizzy within minutes. Then I'll remember that I shouldn't be drinking diet drinks containing aspartame."

TREMORS (SHAKING)
WITH A COMMENTARY ON PARKINSON'S DISEASE

Fifty-one aspartame reactors in this series complained of severe tremors or shaking. The shaking usually represented exaggeration of a disorder known as "benign" or "essential" tremor. The ensuing occupational and social problems associated with this feature, however, were far from benign—as in the case of an engineer who had considerable difficulty controlling his handwriting while using aspartame products.

The occurrence of aspartame-associated tremors sometimes posed a major diagnostic dilemma relative to "early Parkinson's disease." Aspartame reactors in their 40s and 50s anguished over this possibility when they developed a severe tremor. Fortunately, such worry resolved in most instances when the shaking improved promptly following abstention from aspartame.

The apparent interaction between aspartame and several important drugs used to control benign tremor is pertinent. For example, previous successful control with propranolol (Inderal®) occasionally was lost while consuming aspartame products, even with increased dosages of the drug.

Representative Case Reports

An Air Force fighter pilot (Collings 1988) told a Senate hearing that he developed uncontrollable tremors of the left arm after consuming up to one gallon of aspartame-containing drinks daily. He drank these fluids because of marked sweating and thirst in a hot environment— even resorting to carry aspartame drinks on flights. The tremors disappeared when he was stationed in an area where these beverages could not be obtained. The tremors recurred when he resumed aspartame sodas. Four months later, he suffered a grand mal convulsion. There was no further tremor or seizure after abstaining from aspartame products. He was permanently grounded, however, for an "idiopathic partial seizure disorder."

★ ★ ★

A 75-year-old woman developed severe tremors and "no control of my hands, feet and legs" while consuming aspartame-containing soft drinks and chocolate mixes, a tabletop sweetener, presweetened iced tea mixes, and "sugar free" chewing gum. There was concomitant drooling of saliva, "restless legs," decreased vision in both eyes, a severe headache, unsteadiness of the legs, drowsiness, and discomfort on swallowing. She deduced aspartame to be the cause when her complaints subsided after discontinuing such products. She wrote

"I was in and out of the hospital three times from May to July. During this time I was so ill I could not keep food or liquids down. Therefore I had nothing with aspartame. After I was home for a while after the third visit, I realized these conditions had left me. Since then, I will take nothing I might suspect has aspartame in it. It was a very frightening time for me, and I will never chance going through that ever again."

Is There An Aspartame–Parkinson's Disease Connection?

Aspartame or its components and breakdown products (metabolites) may cause changes in the brain comparable to those that appear to initiate or aggravate Parkinsonism. This is especially applicable to dopamine, a neurotransmitter whose concentration is reduced in Parkinsonism. Shabin and Albert (1988) indicated that patients with Parkinsonism ". . . appear to be more prone to aspartame's neurological adverse effects."

A few of the changes in dopamine and serotonin concentrations within the brain induced by aspartame, phenylalanine and aspartic acid are listed.

- Phenylalanine is converted to tyrosine by the enzyme phenylalanine hydroxylase (Chapter 5). Tyrosine undergoes a change to dihydroxyphenylalanine (levodopa, L-DOPA), which then is transformed to dopamine.
- The chronic administration of excess phenylalanine and aspartic acid tends to decrease serotonin and other neurotransmitters within several regions of the brain, and may alter dopamine receptors in certain brain cells.
- Crippling fluctuations in the motor performance of patients with Parkinsonism who are treated with levodopa have been improved by eliminating protein from their breakfast and lunch. Pincus et al (1986, 1987) demonstrated a close correlation between elevated levels of the large neutral amino acids and aggravated symptoms of Parkinsonism, notwithstanding high plasma levodopa concentrations. These motor fluctuations improved when the amino acid levels declined. Such

antagonism of levodopa's action by protein and amino acids probably reflects interference with its transport across the blood–brain barrier (Chapter 5).

SEVERE SLEEPINESS (NARCOLEPSY)

Severe sleepiness not attributable to other causes was a prominent complaint in 17 per cent of aspartame reactors. Some described it as "extreme fatigue" or similar expressions.

- Dozing occurred after chewing as little as one stick of an aspartame-containing gum.
- Several of my patients in their 40s and 50s developed aspartame-associated narcolepsy of such severity that they avoided going out in the evening, and would retire as early as 8 PM.

I have previously elaborated upon the problem of unrecognized *narcolepsy* (Roberts 1964b,c, 1971b). This common condition is characterized by inappropriate and oft-uncontrollable drowsiness and sleep. Its prevalence can be readily inferred from the entrenchment of the "coffee break" and "cola break" in our society—along with the enormous consumption of caffeine and other stimulants. Narcolepsy is *frequently* associated with hypoglycemia (Roberts 1964, 1971).

Aspartame can cause or aggravate severe sleepiness through various mechanisms, especially alterations of norepinephrine and other neurotransmitters (Chapter 5). The precipitation or intensification of narcolepsy by aspartame products among drivers and pilots poses potentially major safety hazards (Chapter 31).

A 51-year-old letter carrier developed "drowsiness, sleep apnea and narcolepsy" after consuming diet colas and an aspartame tabletop sweetener over the previous six months. He also experienced severe dizziness, a marked decrease of vision in one eye, ringing in an ear, memory impairment, "personality changes," nausea, severe thirst and frequent urination at night. A CT scan of the brain, an EEG, lumbar puncture, special eye and ear studies, and a cardiac stress test failed to reveal a definable cause for his symptoms.

The patient rechallenged himself *twice* with aspartame products. His symptoms predictably returned within three days. Although most disappeared within four weeks after stopping aspartame, the tendency to severe drowsiness persisted.

HYPERACTIVITY

Forty-three aspartame reactors complained of hyperactivity of the body, either generalized or limited to their extremities. The latter disorder is referred to as "restless legs." Such continual involuntary movements of the lower limbs are often accompanied by sensations described as numbness, "crawling" or "pins and needles," and by spontaneous leg cramps (Roberts 1965a, 1973).

Children seem more prone to aspartame-associated *generalized hyperactivity*. The Centers for Disease Control (1984) reported the aggravation of hyperactivity in a child when his physician administered aspartame as a part of a double-blind study.

Representative Case Reports

A registered nurse described her aspartame-associated hyperactivity in the following correspondence.

"I become exceedingly restless when I drink a beverage containing aspartame. I have to keep moving for no apparent reason. If I consume anything having this sweetener, no longer can I sit still and read. I have to get up—wander around. The first time this happened to me, I was frankly amazed. I never even realized that it was due to this lo-cal drink."

★ ★ ★

A 55-year-old woman had enjoyed good health until consuming an aspartame iced tea mix in August 1983. She developed "restless leg at night" and aching the following week. These symptoms subsided in the fall when she stopped the tea mix. After resuming it the summer of 1984, however, she reported: "Almost immediately I noticed the same symptoms, and stopped using the product. The aching of the legs ceased."

ATYPICAL PAINS (NEURALGIA, NEUROPATHY)

Many aspartame reactors complained of unexplained pains in various areas. These included the limbs, face, neck, chest and abdomen. Such discomfort often suggested disturbed function either of the peripheral nerves, or of the sensory tracts within the brain and spinal cord. The cause of the pains commonly had defied diagnosis until they improved after the person avoided aspartame products.

Limb Discomfort

Eighty-two aspartame reactors experienced "pins and needles," "tingling," "crawling," "numbness," and "burning" sensations of the limbs—complaints consistent with peripheral nerve irritation. Such sensory neuropathies often are referred to by patients as "neuralgia" and "neuritis."

Pain in the Face

Atypical facial pain was a prominent symptom in 38 aspartame reactors. A number had been misdiagnosed as having temporomandibular joint (TMJ) dysfunction. Significant relief after avoiding aspartame products was crucial to the correct diagnosis.

Dentists in particular must be attuned to this problem. In fact, I initially became aware of its severity after interviewing a dental surgeon whose own facial pain subsided only when he stopped aspartame products. He then uncovered comparable "intractable" facial pain in two patients who were consuming considerable aspartame beverages. Few patients in this category had ever been questioned about aspartame use. Many deduced the association by themselves.

Representative Case Reports

A 42-year-old man with severe pain over the left side of his face had been using two cans of diet cola and six packets of an aspartame tabletop sweetener daily. Numerous studies during two hospitalizations failed to reveal a cause. His pain subsided shortly after avoiding these products.

★ ★ ★

A 44-year-old executive suffered severe facial and head pain for six months that was attributed to the TMJ syndrome. He obtained minimal relief, however, from extensive dental treatments and maneuvers. He also developed blurring of vision in both eyes.

In response to my query about aspartame products, he stated that he was drinking up to three 12-ounce cans of diet soft drinks daily, and chewing five sticks of aspartame gum. His symptoms improved shortly after stopping all aspartame. He then rechallenged himself at least *ten* times . . . with predictable recurrence of the same pain.

MYASTHENIA GRAVIS

Drooping of one or both eyelids (ptosis), difficulty in focusing the eyes, general muscular weakness, and a positive edrophonium (Tensilon®) test characterize myasthenia gravis. These and concomitant complaints occurred in seven persons using aspartame products (Chapter 14). Furthermore, there was remarkable improvement, with or without pyridostigmine (Mestinon®) therapy, after avoiding aspartame. Two patients were able to stop this drug.

SLURRED SPEECH

Prominent slurring of speech—that is, clearly noticeable to others—occurred in 64 aspartame reactors. One can readily sympathize with the concern and frustration of persons in this category whose occupations necessitated personal or phone contact. The problem *promptly* subsided after they avoided aspartame products—and predictably returned following their resumption.

Such impairment of speech could not be ascribed to stroke, alcohol or drugs. It was erroneously attributed to a brain tumor or prior minor head injury in several instances.

Representative Case Reports

A 53-year-old woman developed increasingly slurred speech. She used considerable amounts of aspartame in cereal, coffee, tea and various soft drinks—for example, putting two aspartame sweetener packets into each cup of coffee. Her employer assumed the speech problem was due to the recent false teeth. Other complaints included intense headache, dizziness on sudden motion, and severe depression. She stopped aspartame products after reading an article about aspartame reactions. All symptoms disappeared within three weeks.

★ ★ ★

A housewife was repeatedly kidded by her family for a "bad problem with slurring of my words." She also developed recent impairment of memory and considerable visual difficulty for which no cause could be found. After learning of similar cases attributed to the use of aspartame products, she discontinued diet sodas and an aspartame tabletop sweetener . . . with prompt improvement.

★ ★ ★

A 22-year-old student was hospitalized for a head injury sustained during an automobile accident. He evidenced increasingly slurred speech after leaving the hospital. (At one point, "I could not effectively communicate verbally.") There was concomitant irritability and a striking change of personality. Speech retraining effected minimal improvement.

He had been taking up to 16 cans of diet colas and soft drinks, three packets of an aspartame tabletop sweetener, and one glass of presweetened iced tea daily in an attempt to avoid weight gain. Fortunately, "A friend of my father indicated that some of the speech problems I had been experiencing could be caused by aspartame. My mother sent for more information. Then I quit." The speech and other complaints promptly improved, and were gone by six weeks—only to return within two days after rechallenge. He wrote

"I think that aspartame really had a devastating effect on my speech. Perhaps the depression or irritability was caused because I felt bad concerning the effect it had on my verbal communication. I drank 8 to 16 cans a day for years. Quitting my consumption has had a very favorable effect on my speech and my demeanor."

INSOMNIA

I have been impressed in my practice by the severity of insomnia among aspartame reactors. Most patients promptly improved after avoiding such products. The problem was misdiagnosed as a primary depression in several patients.

Such insomnia was exaggerated by the prescription of hypnotic or tranquilizing drugs capable of altering rapid-eye-movement (REM) activity during sleep (Roberts 1971b). The use of considerable caffeine as coffee or diet colas for aspartame-related "fatigue" also contributed to this vicious cycle.

Representative Case Reports

A young mother related her aspartame-associated insomnia in this correspondence.

"I've found aspartame to cause insomnia and pain in my eyes. I used to live on such products, but when I became pregnant, I stopped them for fear of what it would do to the baby. *All of a sudden, I could sleep at night!* After three months, I started back on some diet sodas, just one a day, and could no longer sleep. I also noticed a sharp pain in my eyes. It was during this time I heard you on the radio program."

★ ★ ★

A male executive suffered severe insomnia when he drank two cans of a diet cola daily. He stated, "The lack of sleep created by it made working next to impossible." The insomnia disappeared one day after avoiding aspartame. It recurred on each of the three occasions he re-tested himself. He added that his wife and son suffered aspartame reactions consisting of "sleepiness, chills, and disorientation."

★ ★ ★

A 57-year-old executive developed severe insomnia. His physician attributed it to "jet lag" because of his frequent travels abroad. He was consuming 10 to 12 cans of an aspartame cola and 4 to 6 packets of an aspartame powdered sweetener daily. A tranquilizer (Halcion®) was prescribed—but only intensified the insomnia. He was referred to a psychiatrist for possible depression . . . then to an endocrinologist for a possible brain tumor. The psychiatrist treated him for *three years,* during which time he was " . . . becoming significantly more moody, nervous, and experiencing memory loss and speech difficulties (frustrating to me, but really not that noticeable to others)." Low doses of the antidepressant drug Tofranil® triggered "a near panic attack." within two weeks after discontinuing aspartame products, he was markedly improved.

CHAPTER THIRTEEN

Psychiatric and Behavioral Reactions

What begins as a trickle of a medical research finding grows as it takes on new ideas and is nourished by corroborative findings. Controversy swirls as the stream encounters boulders of academic dispute, which cloud the original idea. Soon the developing river of awareness takes on new adherents as doubters are enveloped in its flow. In the end, what was once heightened awareness is universally accepted and becomes mainstream.

— DR. JEROME P. EHRLICH
(1987)*

EMOTIONAL AND BEHAVIORAL REACTIONS have been both frequent and severe among aspartame reactors. They may represent the beginning of a new problem or the flare-up (exacerbation) of previous mental disorders.

These adverse effects can be summarized as follows, using one or more of the descriptions appearing in the survey questionnaire (Chapter 7).

* © 1987 Medical Tribune, Inc. Reproduced with permission.

Severe depression	25%
Extreme irritability	23%
Severe anxiety attacks	19%
Marked personality changes	16%
Recent severe insomnia	14%
Severe aggravation of phobias	7%

GENERAL CONSIDERATIONS

Flare-up of Prior Mental Disorders

Astute aspartame reactors often emphasized the recurrence or aggravation of known prior emotional problems. Some examples:

- "The depression while using aspartame was worse than anything before or since."
- "Aspartame adds badness to an already bad problem."

Observations And Insights By Relatives

The "aspartame connection" relative to behavioral changes often was deduced by a spouse or relative.

A 42-year-old woman suffered severe depression with suicidal thoughts while drinking up to four 2-liter bottles of a diet cola daily. She also experienced convulsions, tremors and severe mental confusion. All these complaints disappeared within two weeks after avoiding the beverage. Her husband wrote the following letter:

"My wife has been under the care of a psychiatrist for all her adult life. For the past 5–10 years, her medicine has been primarily Librium® and Stelazine®.

"She has always been a big cola drinker and, being overweight, she opted for diet cola. In 1984, she began to deteriorate physically and mentally. It was gradual. The doctors could not come up with any answers as to the cause. During 1985, she became very bad. In December, she suffered a total collapse. She was hospitalized, and immediately put into the ICU . . .

"When I got her home, I was determined not to let her become re-addicted to the cola, so I would not allow her to have any. She started to show improvement once again. It was about that time that I saw an article in a news magazine that called attention to the fact that aspartame was detrimental to health, particularly to those with nervous conditions . . .

"I am out of town frequently, and several times during the following months she was able to secure the diet cola; each time it made her sick."

Animosity Toward Psychiatrists and Psychologists

Many aspartame reactors took extraordinary offense over what appeared to be the repeated misdiagnoses by personal physicians, psychiatrists and psychologists. Unfounded inferences about "the somatization of deep-seated emotional disturbances" were particularly objectionable.

These patients resented *the indifference of mental-health professionals* when they raised the possible contributory role of aspartame products. Several psychiatrists reflexively rejected this notion because it was "too subjective."

A 31-year-old administrator suffered severe depression, dizziness, three convulsions and marked drowsiness while using four cans of diet cola and up to three packets of an aspartame tabletop sweetener daily. A CT scan of the brain, two electroencephalograms, and other tests were normal. The complaints disappeared within two months after avoiding aspartame. She expressed resentment over having been " . . . told that it was all in my mind, and that it would be wise to consult a psychologist."

Professional Insights

The foregoing experiences have led to my firm conviction that *every patient who presents with unexplained "emotional problems"—recent or recurrent—should be asked about aspartame consumption.* This association is particularly valid under the following circumstances:

- Concomitant neurologic, ocular and medical problems consistent with aspartame reactions.
- Psychiatric deterioration in patients with diabetes or hypo-

glycemia while on previously effective medication and careful adherence to diet.

- The absence of "substance abuse."
- Severe emotional problems and further weight loss in patients with anorexia nervosa, bulimia ("pernicious vomiting"), or variants of these conditions.

Others share my concern. Dr. Ralph G. Walton (1986) urged physicians to consider " . . . the possible impact of aspartame on catecholamine and indolamine metabolism, and inquire about the use of this artificial sweetener when assessing patients with affective [emotional] disorders."

DEPRESSION

Severe depression was a prominent complaint in *one out of four* (139 of 551) aspartame reactors. Its manifestations include insomnia, unexplained fatigue, poor appetite, difficulty in concentration, loss of interest in activities that were formerly enjoyed, a loss of self-esteem, and even thoughts of suicide.

Clinical Considerations

The *prompt onset* of depression after beginning to use aspartame products, especially when it had not been a problem previously, occasionally was dramatic.

A successful businesswoman developed severe depression after drinking aspartame-containing beverages. There was concomitant headache and lethargy. She described her depression in these terms: "I didn't even want to go out into the sun and enjoy the summer. I even looked forward to the rain. (I love summer and hate rain.)" Her complaints promptly receded after abstaining from aspartame products.

A relapse of depression while on previously effective medication occurred with such consumption. Aspartame may interfere with the action of imipramine (an antidepressant) and other drugs that influence neurotransmitters (Chapter 22), probably by decreasing serotonin concentrations in the brain.

Brain turnover of serotonin is decreased in depression and behavioral disturbances, including suicide (*The Lancet* 2:949–950, 1987). Such impairment tends to be reinforced by reactive hypoglycemia and high insulin levels at least among violent offenders and arsonists (Virkkunen 1982, 1983, 1984, 1987).

Suicidal thoughts were admitted by *more than one out of three* persons with aspartame-associated depression. Suicidal ideation was especially impressive in persons attempting to lose considerable weight.
Aspartame-associated depression among close relatives was encountered. The tendency to depression in certain families is well known. Since aspartame reactions often occur among multiple family members, it should not be surprising that such products may initiate or aggravate depression in predisposed relatives.

A woman wrote in the survey questionnaire:

"Yes. I definitely experienced side effects when I use aspartame regularly—depression! I am a happy, upbeat person and never feel 'down' unless I've been using aspartame.

"I have a 28-year-old daughter who suffers from heart palpitations and depression when she uses aspartame. I (also) have a 22-year-old daughter who feels depressed if she uses aspartame."

A 62-year-old woman developed severe depression accompanied by suicidal thoughts. She felt "just miserable, hating every day." There was striking improvement of these and other complaints within six days after stopping an aspartame beverage. Her sister, a grandson and a brother also experienced comparable reactions that promptly improved or disappeared when aspartame was avoided.

Aspartame-associated depression tends to be severe among some older persons. A vicious cycle may occur when the concomitant failure to concentrate and mental blocking are diagnosed as "old age."
Teenagers also are vulnerable to aspartame-associated depression—particularly girls who severely restrict their diets. They may become withdrawn, lose their appetite, and have little interest in school achievement or sports.

Several aspartame reactors experienced *"bipolar depression"*—that is, depression alternating with intense excitement and overactivity (manic behavior).

> I had successfully treated a 67-year-old man for bipolar depression with small doses of lithium over several years. He then experienced "increased mood elevations" shortly after consuming aspartame products. This problem—and associated dizziness—abated after abstinence from aspartame products. They have not recurred for over two years in spite of enormous personal and physical stress incurred by his wife's serious illnesses and multiple hospitalizations.

Public Health Considerations

Severe depression has assumed epidemic proportions in the United States (Holden 1986). The National Institute of Mental Health estimates that six percent of the general population is subject to clinical depression, especially women. Accordingly, *an additive used by more than one-third of the population that could trigger or aggravate depression would pose a potentially enormous public health threat.*

The ongoing decline in the average age—to the mid-20s— when depression begins is noteworthy. This fact is particularly relevant in view of (1) the widespread consumption of aspartame products by young people, (2) an estimated ten percent of teenagers suffering significant depression, and (3) the tripled suicide rate for the 15- to 19-year-old age group. The National Adolescent Student Health Survey of 11,000 eighth- and 10th-graders in 20 states during 1987 revealed that 34 percent(!) reported they had seriously thought about ending their lives. (About 15 percent had actually made a potentially fatal attempt.)

A lawyer perceptively wrote in the questionnaire: "Please research as many suicides as you can regarding *quantity consumed* prior to the act. Also child abuse. Watch in particular the teen suicide situation!"

Representative Case Reports

> A 43-year-old nutritionist developed severe depression within one week after consuming ten or more glasses of an aspartame soft drink.

It subsided when she avoided it . . . only to recur during rechallenge. She wrote:

"I was depressed (sad, crying, etc.) for no reason. I was on vacation at the time and having a good time . . . I started drinking aspartame on vacation because of thirst . . . I do professional counseling. My clients (also) have experienced depression and visual problems."

★ ★ ★

A 43-year-old lawyer quit several jobs because of presumed "job related depression". He drank three or four cans of diet cola daily. He also had suicidal thoughts, extreme irritability, "anxiety attacks," and personality changes. He deduced this aspartame beverage to be the cause after rechallenging himself *four* times. His complaints improved "immediately" every time he avoided it. A niece had aspartame-associated severe depression and menstrual changes.

ANXIETY ATTACKS

One out of five aspartame reactors in this series experienced "severe anxiety attacks." These consisted of generalized anxiety, marked irritability, panic attacks (sweats, rapid heart action, trembling, dizziness, and feeling faint or smothering), phobias, obsessive-compulsive behavior, or combinations thereof.

Representative Case Reports

A 34-year-old nurse developed severe "anxiety attacks" while drinking eight cans of diet cola and chewing ten sticks of aspartame gum daily. She also complained of lightheadedness, "epilepsy-like fits," marked memory loss, slurred speech, headache, sensitivity to noise, diarrhea, severe joint pains, and less-frequent periods. Her symptoms subsided after avoiding aspartame products. She stated:

"My doctor told me I was having 'anxiety attacks.' However, I have not had that feeling since I quit using aspartame . . . I had to convince my physician that aspartame, and not neurosis, was my problem. Luckily for me, my friend developed seizures and linked it to aspartame. I got off it before it happened to me."

Others have reported similar encounters. Drake (1986) described panic attacks in a 33-year-old female who drank considerable amounts of aspartame-sweetened cola drinks.

PERSONALITY CHANGES

Sixteen percent of aspartame reactors evidenced "marked personality changes." Some examples:

- A 34-year-old teacher described her reaction to drinking diet cola as "trouble with co-workers, mood flare-ups, and one or two arguments with the principal of the school even though we had previously got along OK."
- A wife wrote about her "very outgoing and vibrant-on-action" husband becoming an "in-spirit-nothing" person every time he drank an aspartame beverage.
- A 39-year-old woman stated, "I was amazed by the total personality change. My tests were primarily my own. Doctors think I'm 'nuts' to think a sweetener would be the problem." She also developed irritability, memory loss, and insomnia while using aspartame products.

Representative Case Report

A woman became distraught over changes in her personality shortly after using aspartame products. Her greatest concern was "fussing at my husband every minute he was home after we had always had a beautiful relationship . . . I think my husband thought I was psychotic—and frankly, so did I."

She also experienced headache, fatigue, diarrhea, rashes, weakness, a loss of appetite and weight, and "flashes of darkness like I was going to pass out, and jerking in my insides." An antidepressant drug did not help. Aspartame products were stopped when a cousin told her about possible reactions to them. She reported:

"About 24 hours after I took my last diet drink, I suddenly felt better than I had in months. I have been fine ever since. I have continued to stay away from aspartame, and nobody will ever convince me that it was not the cause of all my problems. Also, I credit my cousin for saving my life, because if she hadn't told me about reading that article, I believe it would have killed me."

BEHAVIORAL ABNORMALITIES

Aspartame reactors and their relatives repeatedly described otherwise-unexplained changes in behavior while consuming aspartame products. For example, a young mother "lost my temper and found myself screaming at my son" within one day after drinking a diet cola. This was repeated on the *three* occasions she challenged herself with two different brands.

The potential for aspartame-associated psychopathic behavior should be carefully investigated. This is particularly important in the case of children consuming large amounts who commit murder or other crimes.

CHAPTER FOURTEEN

Eye Complaints

Where there is no vision, the people perish.
— P R O V E R B S (29:18)

IMPAIRMENT OF VISION AND other eye complaints were among the most distressing adverse reactions to aspartame products. Some active adults sensed a threat to their careers and livelihood. Retirees feared the loss of independence if they became unable to drive safely, coupled with concern over the existence of a serious disease (such as glaucoma), the need for powerful glasses, and the prospect of costly consultation or surgery.

The frequency and severity of eye problems in this series of aspartame reactors were striking. Decreased vision was a major complaint in 140 (25.4%), severe pain (one or both eyes) in 51 (9.3%), "dry eyes" or trouble wearing contact lens in 46 (8.3%), and blindness (one or both eyes) in 14 (2.5%).

In my opinion, *specific inquiry should be made about aspartame consumption when an individual develops new or more severe visual symptoms, regardless of the presence of concomitant eye disorders, hypertension or diabetes.* This is particularly true for the increasing number of people with visual changes due to aging that optometrists are encountering.

As with all reactions attributed to aspartame products, the same diagnostic clinical criteria applied to eye problems—namely, improvement after stopping them, and recurrence on resumed use. Of particular concern to most reactors with eye symptoms was the possibility that severe impairment or even loss of sight might not diminish appreciably after abstinence from aspartame.

Other Diagnostic Considerations

In most of these patients, there was no convincing evidence for underlying glaucoma, occlusion of a retinal vessel, toxic amblyopia (related to excessive alcohol or smoking), or optic neuritis due to multiple sclerosis and other causes that might account for the symptoms. CT scans and MRI studies of the brain or optic nerves generally proved normal in these patients.

Furthermore, that patients had known cataracts, astigmatism, macular degeneration or diabetic retinopathy did not necessarily disprove the role of aspartame . . . especially when vision *promptly* improved after stopping aspartame products. In other words, part of the problem may have been due to aspartame or its components. This situation occurred in several instances despite futile surgery.

> The visual problems in a 72-year-old woman persisted after a cataract had been removed. She also complained of severe headaches, lip–tongue pain and swelling, nausea, and "nervousness." All symptoms subsided when she then avoided aspartame products.

> A 30-year-old female executive had a radial keratotomy performed on the left eye because "my eyes would intermittently go out of focus" while drinking three cans of diet soda daily. A *second* keratotomy was done on the same eye six months later due to persistent visual complaints. I subsequently saw her in consultation for other aspartame-related complaints and concluded that the aspartame product probably contributed to her persistent problem.

Needless or premature cataract extraction in diabetics whose visual symptoms largely represent a reaction to aspartame products poses a potential added problem: the progression of diabetic retinopathy (Jaffe 1988).

A recurrent paradox was encountered—the inability of some aspartame reactors to convince ophthalmologists about the legitimacy of their visual symptoms. Many were told, "Your eyes are fine. You don't even need new glasses."

Some patients consulted me *after* repeated visits to eye doctors for continuing, but unexplained, visual difficulty. The "aspartame connection" surfaced when they related other complaints commonly experienced by aspartame reactors—especially severe headache and confusion. As noted earlier, however, improvement of their sight tended to be delayed longer than other symptoms after abstinence from aspartame products.

A 56-year-old engineer had been successfully treated by me for reactive hypoglycemia and hypothyroidism with appropriate diet and medication. He returned because of progressive difficulty in vision for several months that could not be explained by his ophthalmologist. He also was having severe headaches, confusion and memory loss (of particular concern in his professional work), "dry eyes," marked dizziness, tingling and numbness of the limbs, depression, episodic shortness of breath, abdominal pain with diarrhea (occasionally bloody), intermittent difficulty in swallowing, intense thirst, and frequency of urination (both day and night.) Severe drowsiness virtually precluded his going out in the evening, causing him to retire at about 9 PM. Specific inquiry revealed that he had been using four cans of an aspartame cola, up to two bowls of an aspartame cereal, and other aspartame products daily. He then volunteered that his daughter also had aspartame-related visual problems. The headaches improved considerably one week after abstaining from such products. His clarity of thinking and other symptoms improved two weeks later, along with definite improvement of vision.

High-Risk Groups

Certain categories of persons appear more vulnerable to aspartame-related visual problems. They include older individuals, patients having diabetes (with or without overt diabetic retinopathy), severe hypoglycemia, and those who consume much alcohol and tobacco.

A 51-year-old insulin-dependent diabetic had been using three cans of diet sodas and eight packets of an aspartame tabletop sweetener daily for three months when she noticed strange visual images. "I saw things, like a puppy in my kitchen, and we don't have a dog. Also shadows flitting by in my hallway." Other recent complaints included headache, lightheadedness, "pins and needles" in the limbs, palpitations, severe depression and marked irritability. All symptoms disappeared within several days after avoiding aspartame products.

Pathophysiologic Influences

There are several possible explanations for eye symptoms in aspartame reactors. The high energy requirements of the retina (Roberts 1971b) render it uniquely vulnerable.

- Methanol (Chapter 6) can cause blindness as a result of edema of the optic disc (Hayreh 1977) and degeneration of the retinal ganglion cells (Baumbach 1977). In addition to optic atrophy, patients with methanol intoxication have developed Parkinsonism, dementia, and other neurologic abnormalities (McLean 1980).
- Changes in the retina have been induced in young rats with seven percent phenylalanine (Dolan 1966).
- The intake of considerable amounts of the D-aspartic acid isomer present in aspartame products, especially as a result of excessive heat or prolonged storage (Chapters 5), may influence vision. In this regard, there is progressive accumulation of D-aspartic acid in the lens nucleus with aging (Man 1983).

VISUAL COMPLAINTS

The spectrum of aspartame-associated eye symptoms includes blurred vision, "double vision," "tunnel vision," "bright flashes," "eye pain," intense sensitivity to glare, repetitious "jerking of the eyes" (nystagmus), repeated blinking "to focus my eyes," recurrent "bulging of my eyes," and total or near-total loss of vision in one or both eyes. A 35-year-old anesthetist told a Senate hearing: "I began having visual disturbances making it difficult to see the monitors in surgery and to fill out my anesthesia record" (Taylor 1987).

A 36-year-old woman experienced recurrent blurring of vision, "unusual lights and floaters," and pain in both eyes while drinking two 2-liter bottles of an aspartame orange drink, and up to five glasses of presweetened ice tea daily. She also suffered severe headache, marked memory loss, mental confusion, slurred speech, intense depression, abdominal bloat and severe itching. She neither smoked nor drank alcohol. Her complaints abated after avoiding these products.

A 34-year-old woman described her ocular reaction to aspartame products as follows:

"Whenever I drink an aspartame pop, my eyes cloud and I get a mucus covering. I know it is from the pop because I won't have it for a week. Then on the day I have it, my eyes are thick with a mucus covering by night."

A 50-year-old woman experienced a marked decreased of vision in both eyes while drinking diet colas and aspartame-sweetened tea. She complained, "I needed a prescription (for glasses) for the first time in my life!" She also developed tingling of the hands, slurred speech, palpitations, and a symmetric rash. All these features improved within several days after avoiding aspartame products.

A 37-year-old businessman consumed three to four glasses of aspartame beverages daily for one year prior to the onset of severe headaches, impaired vision in the right eye, and severe thirst. He wrote me this account of subsequent events:

"My problem started just prior to October, 1985. I had been having trouble with my vision and getting headaches, so I went to see my ophthalmologist. He was not able to find anything wrong and suggested that I take some time off. I didn't and just lived with the problem until the summer of 1986. Toward the end of the summer, the headaches became worse to the point that I was unable to function, and had to go to sleep for several hours. Also, I started to lose the vision in my right eye. I returned to the ophthalmologist about the beginning of August. He found during the examination that I had severe hemorrhaging in my retina but couldn't explain why. He recommended that I see a retinologist. He too was unable to explain what was causing my problem, and recommended that I see a neuroretinologist at the same hospital. I should mention that I do not drink (alcohol) very often.

"I went to see this doctor because of the headaches mentioned. He put me into the hospital for a full range of tests including a CAT scan from head to toe, a full series of blood tests, and spinal taps. With the exception of seeing some cells in my retinal and spinal fluid, everything came back negative. I continued to see the retinologist who started me on steroids. After the hospital, my vision didn't get any worse nor did it get any better. The headaches stopped and my retina seemed to be healing.

"Coincidentally, at about the time I went into the hospital, my wife read an article in the newspaper that cautioned against serving drinks with aspartame to children. She stopped making it, and obviously I stopped drinking it because it wasn't around. I didn't make the connection between the aspartame and my medical problems until about two weeks ago when I found a container of an aspartame soft drink and drank about two glasses. Within minutes I started to get the headache again, my vision became blurry, and the skin on my skull started to hurt. I discussed the incident with my wife, and we put two and two together.

"Sadly, there was no reason that my vision should have deteriorated as much as it has except that the doctors I saw didn't know the right questions to ask. It would seem that not enough is known about aspartame to allow it to be used without some kind of warning."

A 45-year-old traffic controller experienced intense pain and blurring of vision in one eye, along with "dry eyes," three days after drinking two 10-ounce bottles of a diet cola daily. He wrote, "The pain was so bad it doubled me up and I could hardly move. I forced the eye open with both hands and let water spray into my eye. My eye hurt all morning and was red for 24 hours."

Several days later, his vision was clouded on awakening by a "road map of black blood vessels, which lasted about three seconds after each time I blinked. This condition went away after four or five minutes." Other symptoms included intense thirst and severe "low blood sugar attacks." The "road map" and eye pain disappeared within two days after he avoided aspartame. He retested himself *three times,* however, before concluding that the aspartame drinks had caused the problem.

A 55-year-old priest found it difficult to perform his duties because of impaired reading and intermittent double vision. He had been using three cans of aspartame-containing soft drinks and three packets of an aspartame tabletop sweetener daily. Concomitant complaints included ringing in the ears, confusion, slurred speech, severe headache, itching and joint pains. All disappeared within four days after avoiding aspartame products—only to predictably return within three days after rechallenge.

Loss of Vision

Fourteen persons in this series reported extreme loss of vision or frank blindness in one or both eyes. It was transient in most, but permanent in a few. No other convincing cause for the sight loss could be found by multiple consultants over prolonged observation . . . during which time the patients had continued using aspartame products.

The *FDA* has received reports on two persons with *total bilateral blindness* (personal communication, Dr. Linda Tollefson, August 12, 1987). One was a nine-month-old permanently blinded child whose mother drank two diet drinks daily during pregnancy. The other was a female with severe optic neuritis that eventually responded to corticosteroid therapy.

A 41-year-old woman noted severe visual loss after drinking aspartame. She stated, "If I drink two or three diet colas a day, I lose vision in my right eye for 10–15 minutes a session.

A 37-year-old woman drank four to five cups of coffee with *two* packets of an aspartame tabletop sweetener in each cup every morning, one or two cups of similarly sweetened coffee in the mid-morning, one diet cola, and up to six glasses of aspartame-sweetened iced tea daily (especially during the summer) for 18 months. She developed a constant headache, dizziness, insomnia, fatigue, depression and memory loss. There was subsequent profound loss of vision in the left eye on awakening one morning; this progressed to complete blindness within four days. Extensive medical, neurologic and ophthalmologic consultations failed to uncover a convincing cause for the "optic neuritis." Steroid (prednisone) therapy proved minimally effective.

She discontinued these aspartame products after discovering another person who attributed a similar problem to this sweetener. Her headaches, weakness and insomnia abated within one month, but the visual loss persisted several years—that is, only slight sensitivity to light being present in the affected eye. She could not resume her prior occupation as a jewelry artisan due to severe impairment of depth perception.

"Dry Eyes" And Difficulty With Contact Lens

Forty-six aspartame reactors (8.3 per cent) in this series experienced "dry eyes," irritation from contact lens, or both. There was generally prompt improvement after abstinence from aspartame products.

A 47-year-old woman stated that aspartame products caused marked dryness of the eyes, which also precluded wearing contact lens. She required one bottle of artificial tears a week, but was able to discontinue their use after avoiding such products.

The frequency of concomitant "dry mouth," memory loss, and psychologic changes in these aspartame reactors may have some relevance to the Sjögren syndrome. This latter disorder affects about two percent of adults. It is characterized by reduced or absent secretions from the tear (lacrimal) and salivary glands. As many as one-fourth of patients with Sjögren's syndrome have cognitive impairment ranging from forgetfulness to severe memory loss.

Precipitation of Clinical Myasthenia Gravis

Myasthenia gravis (Chapter 12) is a form of generalized muscular weakness. One or both upper eyelids usually droop (ptosis). The diagnosis of this disorder is confirmed by prompt improvement after giving edrophonium chloride (Tensilon®) intravenously. Treatment consists of pyridostigmine bromide (Mestinon®) and other measures.

I have personally studied three female aspartame reactors in whom recent drooping of the eyelids and marked muscle weakness were prominent features. The diagnosis of myasthenia gravis had never been raised during at least a decade of attending each patient. All experienced improvement of other aspartame-related complaints when they avoided such products . . . but not the ptosis or severe fatigue. Each evidenced a positive diagnostic response to intravenous edrophonium, and subsequent improvement on pyridostigmine. (This drug could be stopped in two of the patients after they avoided aspartame products.)

An 82-year-old insulin-dependent diabetic female presented with drooping of the right eyelid, which she had to manually elevate. Her ophthalmologist recommended surgery "to pull up the skin."

She was drinking large amounts of aspartame products. Related complaints included memory loss, confusion, dizziness, insomnia, and unexplained deterioration of diabetes control. Her problems improved within one week after avoiding aspartame products except for the drooping eyelid.

There was a dramatic response to edrophonium testing. Pyridostigmine afforded gratifying and persistent improvement. A brief attempt to discontinue it, however, resulted in the prompt return of ptosis.

Four additional cases of diagnosed or suspect myasthenia gravis surfaced among aspartame reactors who completed the survey questionnaire. (Unfortunately, the initial survey did not specifically ask about this diagnosis or its symptoms.) Two with eyelid ptosis were virtually unable to lift themselves out of bed in the morning. The ptosis and generalized weakness in the other two had been attributed to "aging."

This small group of patients raises the intriguing possibility that aspartame may precipitate or aggravate latent myasthenia gravis.

CORROBORATION

Ophthalmologists and other professionals have told me about dramatic improvement of vision in their patients after the cessation of aspartame products.

Dr. Morgan Raiford, an Atlanta ophthalmologist with longstanding interest in the ocular complications of methanol, had encountered 65 aspartame users who presented with severe unilateral or bilateral visual deterioration and/or optic nerve atrophy as of July 10, 1986 (personal communication).

Dr. Woodrow C. Monte, Chairman of the Food Sciences Department at Arizona State University, informed me that up to 60 per cent of 1,300 alleged aspartame reactors who contacted him had ocular complaints. Their symptoms included blurred vision, bright flashes, tunnel vision and blindness.

The FDA has received complaints from 177 consumers who attributed their visual changes to aspartame products. The data, kindly supplied by Dr. Linda Tollefson (August 12, 1987), appear in Table 14-1.

Table 14-1. *Aspartame-Associated Visual Problems FDA Data—177 Cases*

Sex		
Females	142	(80%)
Males	35	(20%)
Race		
White	72	(41%)
Nonwhite	5	(3%)
unknown	100	(56%)
Age groups (years)		
1–19	3	(2%)
20–29	18	(10%)
30–39	33	(19%)
40–49	36	(20%)
50–59	18	(10%)
60 +	32	(18%)
Unknown	37	(21%)
Aspartame Products Consumed		
Soft Drinks	119	(67%)
Tabletop Sweeteners	83	(47%)
Hot Chocolate	17	(10%)
Iced Tea	17	(10%)
Gum	7	(4%)
Soft Drink Mixes	32	(18%)
Puddings/Gelatins	32	(18%)
Concomitant Symptoms		
Headaches	79	(45%)
Depression/Mood Changes	30	(19%)
Numbness/Tingling/ Other Neurologic	29	(16%)
Dizziness	64	(36%)
Abdominal Pain/ GI Complaints	29	(16%)
Memory Loss	20	(11%)
Weakness/Fatigue	12	(7%)

Table 14-1. *Aspartame-Associated Visual Problems FDA Data—177 Cases—continued*

Visual Fainting	4	(2%)
Hives/Rashes	8	(5%)
Sleep Problems	13	(7%)
Heart	4	(2%)
Bone/Joint Pain	3	(2%)
Menstrual	4	(2%)
Breathing Difficulty	3	(2%)
Coma	4	(2%)

CHAPTER FIFTEEN

Ear Complaints

How many ears must one man have
Before he can hear people cry?
— *B O B D Y L A N*

IN THIS SERIES OF 551 aspartame reactors, 73 (13%) complained of "ringing" or "buzzing" of the ears (tinnitus), 47 (9%) of marked intolerance to noise, and 25 (5%) of significant hearing impairment. These symptoms often were severe. The tinnitus was described by some as hissing, humming, whistling, pulsating, or like the sound of crickets or a high-tension wire. Patients also conveyed their symptoms as:

- "hearing problems in conversation"
- "as if my ears were covered"
- "severe pressure in my head (that required) equalizing the pressure by blowing my ear drums and holding my nose"
- "random fade-outs in my hearing"

Some reactors were able to pinpoint the onset of their ear symptoms to *the specific day or week* in which they *first* used an aspartame-containing product.

Difficulty in Diagnosis

Problems in recognizing these aspartame-associated ear complaints were the rule. Diagnoses included "aging" (even among persons in their 40s and 50s), Meniere's disease, and suspected tumors of the acoustic nerve or the brain. CT scans, MRI studies, EEGs, and other tests usually were normal.

The failure to consider reactions to aspartame products is underscored by an evaluation of 13,000 tinnitus sufferers by the American Tinnitus Association. Under the likely cause of their distress, more than one-third indicated "nothing known."

Concomitant Symptoms

The majority had other complaints attributable to aspartame products—especially severe dizziness, headache and various neurologic or psychiatric features. Use of considerable amounts of aspirin to obtain relief of concomitant headache or joint pain probably compounded the tinnitus and hearing loss in some.

Representative Case Reports

A 56-year-old man with hypertension responded well to salt restriction and minimal blood pressure medication. He returned to my office with progressive "ringing" in both ears, confusion, insomnia and "nervousness" during recent weeks. The physical examination proved normal. He had begun drinking aspartame beverages the preceding month. His symptoms disappeared within three days after avoiding such products. They have not recurred over the ensuing seven months of observation.

★ ★ ★

A 30-year-old woman consumed an average of five to six cups or glasses of aspartame beverages daily for 17 months. Ringing and pain in both ears, dizziness, and a severe headache began after the tenth month. Audiometric studies revealed considerable loss of hearing in the left ear. When brain tumor was ruled out by CT scan, the otology and neurology consultants diagnosed her problem as Meniere's disease. The patient herself, however, finally deduced that the aspartame drinks caused these symptoms because she could *predictably* reproduce them on rechallenge.

★ ★ ★

A 36-year-old woman complained of "buzzing in the ears." There was

concomitant disorientation and a feeling "like a drug or alcohol muf-
fling the senses." She had been placing one packet of an aspartame
tabletop sweetener into each of eight cups of hot tea daily. She also
experienced marked memory loss, unsteadiness, drowsiness, depression, and severe nausea. Her symptoms promptly improved after
avoiding aspartame products.

CHAPTER SIXTEEN

Excessive Weight Loss: Anorexia and Bulimia

The belief in absurdities is bound to lead to the commission of atrocities.

— V O L T A I R E

PERSONS WHO CONSUME ASPARTAME-containing products often do so to lose or "control" their weight. In fact, nearly half of the aspartame reactors who completed my questionnaire indicated that weight control was the primary reason for using these products.

Loss of appetite (anorexia) and excessive weight loss were prominent features in 26 aspartame reactors. The latter ranged up to 80 pounds—far in excess of the weight loss expected by replacing sugar intake or diet. Put differently, some aspartame users lost huge amounts of weight apparently as reactions to aspartame itself.

The combination of severe weight loss, intense thirst and frequent urination often caused attending doctors first to suspect underlying diabetes mellitus.

Some individuals persisted in consuming aspartame products after being urged to avoid them, however, because of slight gain in weight after their cessation. One cachectic-appearing young businesswoman told me, "I'm *still* too fat."

REPRESENTATIVE CASE REPORTS

A 30-year-old licensed practical nurse drank up to six cans of diet cola daily for six months. She lost 80 pounds (to a weight of 98). The diagnoses of anorexia nervosa and "pernicious vomiting" were made.

Other complaints included severe nausea and diarrhea, decreased vision in both eyes, recent "dry eyes," ringing in both ears, severe headaches, dizziness, petit mal attacks, drowsiness, mental confusion and memory loss, slurred speech, severe depression (with attempted suicide on two occasions), marked irritability and anxiety, palpitations, intense itching, a rash, extreme thirst, frequent urination, joint pains, and "low blood sugar attacks."

This nurse was treated by "at least 10 different doctors with at least 10 different diagnoses." She asserted, "All of my problems began with drinking diet cola."

★ ★ ★

A 29-year-old woman lost 80 pounds over six months. She drank "enormous amounts" of diet cola, and no longer could eat solid food. Many physicians had been seen for associated problems—including "eye doctors, dentists, ear-eye-nose and throat specialists, psychiatrists, and at present an endocrinologist." They included severe headache, deterioration of vision, dizziness, personality changes, memory loss, extreme fatigue, and stomach pains.

RELATED CONSIDERATIONS

Physiological Perspectives

The physiology of weight is complex. The modest "obesity" that psychologically troubles many persons in our society probably serves as a valuable reserve of energy for the body (Roberts 1964, 1965, 1971b). In fact, the insulin-related "thrifty gene" favoring deposition of fat constitutes a survival trait evolved over millennia in response to famine, fasting and hypoglycemia (Chapter 17).

Aspartame can affect the brain's control of appetite (Blundell 1986). Considerable evidence suggests that it alters neurotransmitter function within the satiety center of the hypothalamus . . . in turn leading to anorexia and weight loss (Rogers 1973, Ceci 1986). This may be due to changes involving serotonin and corticotropin-releasing hormone that inhibit feeding, and an increased sensitivity to stimuli satiating the appetite.

An additional consideration is the apparent lessened digestibility of proteins containing excessive amounts of the D-phenylalanine and D-aspartic acid stereoisomers (Chapter 5). This not only could interfere with the transport and metabolism of other amino acids, but also account in part for severe weight loss among aspartame sufferees.

Medical Complications

Aspartame-related weight loss can contribute to some of the serious complications noted after arbitrarily excessive dieting and strenuous exercise (Roberts 1985). These so-called "slimmer's diseases" include *abnormal heart rhythms* (notably ventricular tachycardia), *pulmonary embolism* (clots to the lung), *depression, headache, multiple sclerosis,* and *adverse fetal effects.*

The Role of Contemporary Attitudes

Few would deny the legitimate desire to avoid gross obesity. Yet, most patients in this category of reactors to aspartame products were weight-conscious women afflicted with the "fear of fat." They used them as a "culinary bypass" in conjunction with extreme caloric restriction and exercise.

Unfortunately, even marginally "overweight" persons find themselves trapped in our image-oriented culture. This stems from peer pressure and extreme guilt on viewing the thin figures of young men and women in ads promoting aspartame products. The extraordinary weight loss by three female movie or television superstars has had great impact upon other women who worship "at the shrine of perpetual lettuce," a term used by Erma Bombeck (1988). The implication is loud and clear: "Lose weight and join the fun crowd!"

In point of fact, however, individuals with the "five-pound overweight neurosis" (a term used by columnist Ellen Goodman) often eat much less than their normal-weight peers.

Other influences exacerbate the matter. They include:

- The emphasis by priests of high-priced fashion on emaciation and "angularity," even though most men find natural curvaciousness appealing.

- The attitude held by some mothers and friends: "You can't be too rich or too thin."
- Misdirected journalistic emphasis that equates thinness with "health."
- The barrage of sophisticated advertising campaigns promoting "lite" products for "weight control."
- The "party line" paranoia of some physicians who are satisfied only with the attainment of "ideal weight"—even for individuals in otherwise good health who have maintained a slight degree of "excess" weight for prolonged periods.

The bottom line: an estimated six or seven out of every ten American women dislike themselves for being "overweight." Indeed, one out of three *nine-year-olds* actively worries about becoming fat!

Medical and Public Health Considerations

This preoccupation with thinness is a national disgrace—at least in my opinion. It requires a reorientation of the public, particularly since anorexia nervosa and bulimia (pathologic vomiting, the binge-purge syndrome) are reaching near-epidemic levels. These disorders afflict from 10 to 15 percent of adolescent girls and young females in the United States (Health & Public Policy Committee 1986), and one out of five college women! (They also affect young men, but not so frequently.) *It is my belief that aspartame should not be used by patients with anorexia nervosa and bulimia, pending convincing proof it does not contribute to the problem.*

Physicians and social scientists must consider some of the serious implications of this reversed attitude to female "natural curves," especially as it relates to the optimum bearing of children. For instance, the long-appealing "Venus de Milo look" is now regarded with disdain as obesity by many contemporary women.

There are legal connotations as well. Numerous courts have ruled that manufacturers and producers have a "post-sale duty to warn" when certain products may be "unreasonably unsafe" . . . at least for a segment of the exposed population. They are held to the current status of knowledge possessed by investigators when newer complications, not foreseen in the pre-licensing phase, are encountered—as in the case of aspartame-associated severe anorexia and weight loss (Roberts 1986, 1987a, 1988e). Furthermore, appropriate warnings must be incorporated into labeling, using understandable language for the ordinary consumer, notwithstanding the ability of corporate defense experts to argue about the absence of "absolute proof" (Chapters 30 and 31).

CHAPTER SEVENTEEN

Paradoxical Weight Gain

This is what we must learn in relation to helping others. Because we care, we would like to spare them our misfortune by warning them to beware. However, owing to human nature, rarely is this possible. Usually our advice is scorned even to the point of ruining our friendship if we should persist.

— *T. S. ELIOT*

THIRTY-FOUR ASPARTAME REACTORS in this series *gained* weight while consuming aspartame products. The average weight gain was 19 pounds. The term "paradoxical" indicates that aspartame products were used with the specific intent to lose or "control" weight, when in fact the reverse occurred.

Representative Case Report

A successful businesswoman unexpectedly gained 50 pounds while drinking aspartame beverages. She also developed fluid retention, frequency and discomfort of urination (day and night), severe fatigue, and depression.

"I drank diet soda for the obvious reason—to avoid sugar and to avoid weight gain. The interesting thing is that I remember often thinking, 'My body knows that it is not sugar . . . it's not fooled,' and I would go looking for the real thing. In other words, it did not accomplish my goal of finding a sugar substitute. I cannot be sure it increased my craving for sugar . . . No matter how much I exercised, eating light, etc., I could not lose even a pound. All of the foregoing symptoms disappeared after discontinuing aspartame."

Physiologic Mechanisms

Several mechanisms may contribute to such paradoxical weight gain with aspartame products.

- The addition of an intense sweetening agent to low-calorie foods confuses the satiety center, thereby creating a "reservoir" of residual hunger (Blundell 1986).
- The taste of sweet initiates a nerve-mediated reflex that stimulates insulin release. If the expected energy input (namely, food) is not forthcoming, hypoglycemia—and its attendant hunger—may ensue.
- In anticipation of food, aspartame and phenylalanine can trigger such reflexive "cephalic responses" that result in the release of insulin, cholecystokinin, and various neurotransmitters.
- Concomitant "forced feeding"—that is, eating only one meal a day—favors obesity and related metabolic changes (Bortz 1969).
- The conversion of phenylalanine to tyrosine may be impaired in obese persons (Brown 1973), thereby creating another possible vicious cycle involving increased phenylalanine retention after aspartame consumption. (Fasting plasma concentrations of phenylalanine already tend to be significantly higher in obese persons [Caballero 1987].)

RELATED CONSIDERATIONS AND PERSPECTIVES

Metabolic Defense

Aspartame-associated weight gain can be viewed as an inherent metabolic defense mechanism by "the wisdom of the body" when it is threatened by caloric deprivation. Even relatively small amounts of aspartame products appear to be capable of initiating this biological hunger response.

A 37-year-old woman developed irritability and a change in personality while consuming only one can of diet cola daily. She emphasized, "I was extremely hungry while drinking the soda."

I suggest that physicians, dietitians and others involved in helping compulsive overeaters should encourage them to avoid or minimize the use of aspartame products—notwithstanding the barrage of ads alleging or inferring their effectiveness in "weight control."

This is not a casual assertion. My own research on obesity and weight reduction extend more than three decades, including original studies on Metrecal® (Roberts 1960, 1962). They encompass the contributory roles of hypoglycemia, narcolepsy (pathologic drowsiness), and forced feeding (Roberts 1964, 1965, 1971b).

Insulin is necessary for storing fat. Accordingly, the "hyperinsulinized state" remains a central consideration in understanding obesity. This is especially relevant for individuals with severe reactive hypoglycemia ("low blood sugar attacks") who evidence proneness to diabetes ("decreased glucose tolerance") (Roberts 1964, 1965, 1971b). Such individuals intuitively react to the increased hunger and thirst induced by excessive insulin secretion by consuming more sugar, simple carbohydrates and anything sweet.

The Dubious Advantage of Sweeteners

The use of appreciable amounts of aspartame, in conjunction with drastic caloric restriction, tends to alter the perception of both taste and satiety. This may enhance the consumption of excessive fluids, sugar and fat over the long term . . . a phenomenon termed "the slender trap."

The majority of overweight patients whom I see in consultation are not aware of the possible contributory role of large amounts of aspartame products, saccharin and other sweeteners (Chapter 35) in provoking the vicious cycles mentioned above that are conducive to excessive weight. Consequently, such persons should avoid aspartame and other sweeteners. Furthermore, an accumulating literature challenges the long-term benefits of aspartame products and saccharin relative to sustaining weight loss.

- The American Cancer Society (1986) documented the fact that persons using artificial sweeteners gain more weight than those who avoid them.
- The ingestion of saccharin and other sweeteners in drinking water by experimental animals induces greater food consumption and obesity (Merkel 1979).
- Farkas and Forbes (1965) found that the use of non-caloric sweeteners does not reliably influence "adherence to a carbohydrate-restricted diet by patients with diabetes in the age range of forty to seventy years."

CHAPTER EIGHTEEN

Allergic Reactions

It is hard to estimate how many people who already have allergies are suffering from food additives.

— DR. WILLIAM C. GRATER
(1969)

STRIKING ALLERGIC OR HYPERSENSITIVITY reactions to aspartame-containing products have been reported. They include the following among this series of 551 apparent aspartame reactors:

Severe itching without a rash	8%
Severe lip and mouth reactions	5%
Hives (urticaria)	5%
Other skin eruptions	9%
Aggravation of respiratory allergies involving the nose, sinuses and bronchial tubes	2%

In addition, individual patients developed impressive reactions involving the skin and hair while ingesting aspartame products. They consisted of marked loss or thinning of the hair (31 cases), acne vulgaris ("bad complexion"), acne rosacea, and eruptions resembling a serious disease known as lupus erythematosus.

Other reactions, mentioned elsewhere in this book, also may have an allergic or so-called autoimmune basis. Examples are myasthenia gravis, swelling of the salivary glands, and joint problems.

GENERAL OBSERVATIONS

Female Preponderance

A preponderance of females among patients with allergic reactions to aspartame products was clearly evident. Allergies to food and drugs are known to occur more frequently in women (McLain 1986).

Prior History of Allergies

Allergies to dust, pollens, molds, foods, drugs and other additives existed in one out of three aspartame reactors.

One aspartame reactor with a severe rash reported, "I (also) am allergic to medications and many foods with MSG, sulfites, additives, chocolate, many oils, oranges, (and) milk."

A past history or family background of allergies was common in these individuals.

Aggravation of Existing Allergies

Respiratory allergies seemed to be aggravated by aspartame products. This should be of particular concern for child consumers in light of the apparent increased frequency and severity of asthma among younger persons during recent years.

Ensuing Allergies

Some patients with reactions to aspartame products subsequently evidenced severe allergies to other additives, foods and drugs. They included milk, monosodium glutamate (MSG), corn syrup (especially in a "non-dairy coffee sweetener"), aspirin, and nonsteroidal anti-inflammatory analgesic drugs (such as ibuprofen).

The cited allergic manifestations may represent responses to the aspartame molecule, its components, or the diketopiperazine (DKP) metabolite (Chapter 5). The possible contributory mechanisms involve technical terms and concepts that are not appropriately detailed in this book.

ITCHING, HIVES AND OTHER RASHES

Severe itching and various eruptions occurred in *one out of five* (20%) of aspartame reactors. This exceeds by far the 2.2 percent incidence of drug-induced skin reactions reported by the Boston Collaborative Drug Surveillance Program (Bigby 1986). Accordingly, the role of aspartame products must be *specifically* considered in patients who present with unexplained rashes, especially chronic and recurrent hives (urticaria).

Biological Threshold

Some reactors developed an itch or rash *only* when their daily intake of aspartame exceeded a basic or threshold amount. In virtually all instances, this was less than the maximum acceptable daily intake recommended by the FDA (Chapter 32).

A 34-year-old woman found that her severe itching and eruption did not occur if she limited the intake of diet cola to fewer than four glasses a week. They, however, predictably returned after ingesting four or more glasses a day.

Representative Case Reports

A 52-year-old woman developed itching and hives while consuming six packets of an aspartame tabletop sweetener, three glasses of pre-sweetened iced tea, two glasses of aspartame hot chocolate, and two servings of aspartame-containing puddings or gelatins daily in an attempt to lose weight. Other debilitating complaints included headaches, dizziness, drowsiness, memory loss, decreased vision and pain in both eyes, recent "dry eyes," slurred speech, severe depression, marked change in personality, abdominal bloat, and sensitivity to noise. Her symptoms subsided after stopping aspartame products, but recurred within eight hours on the *two* occasions she rechallenged herself.

Although there had been no previous history of allergies, reactions to other products ensued. She reported, "Today, after two years of being off aspartame, I notice sensitivity to all food additives and other environmental toxins."

★ ★ ★

A young woman developed an extensive rash whenever she ingested products containing aspartame. She then used saccharin-containing beverages without incident. When she drank a cola beverage in which the formulation had recently substituted aspartame for saccharin, an intense itch promptly occurred.

★ ★ ★

A woman timed her one-and-only reaction to an aspartame beverage with precision.

"On April 16, 1984, I had my first glass of diet cola sweetened with aspartame. Approximately one-half hour later, I experienced generalized itching with the eruption of giant hives (urticaria) on my back, chest and arms. There were approximately one hundred total. I never had such an experience . . . in approximately 3 hours, the eruption was alleviated with an antihistamine and cortisone."

★ ★ ★

A 49-year-old female was "miserable and discouraged by the continual outbreaks of hives for one year." This tennis professional also suffered swelling of the lips and eyes, and heavy menstrual bleeding. Three physicians and consultants were unable to find the cause, notwithstanding extensive allergy testing.

The patient then chanced to view a program on aspartame reactions. She recalled that her rash first appeared several weeks after using a "sugar-free" gelatin. The hives did not recur when she avoided such products.

★ ★ ★

A 43-year-old man had recurrent hives over most of his body for two months. They lasted from 20 minutes to eight hours. Numerous tests failed to reveal the cause. He and his wife then engaged in a process of "one-by-one elimination." The rash disappeared when he stopped aspartame-containing beverages. Further proof came during an inadvertent challenge. He wrote

"While on vacation in British Columbia several weeks ago, I bought what I thought were two regular soft drinks out of a machine. Within twenty minutes of drinking them, I began to break out and itch. Within eight hours, I was completely covered with wheals all over my upper torso and neck."

★ ★ ★

A 29-year-old female chemist developed swelling of the lips, tongue, eyes, hands and feet within less than one hour after ingesting an aspartame beverage. It predictably recurred on more than 20 rechallenges. Hives at the age of eleven had been attributed to chocolate. Other allergies ensued after her aspartame reactions. She wrote, "Then suddenly, milk products started causing the same reactions . . . I have never been allergic to milk before."

Observations of Others

As of early 1987, the FDA received 111 volunteered complaints of rashes and 80 complaints of hives attributed to aspartame products by consumers (Tollefson 1987).

After his initial report on this subject, Dr. A. Kulczycki, Jr. (1987) reported that 224 patients had contacted him for chronic hives or severe generalized edema (angioedema) purportedly due to aspartame sensitivity. He estimates that a minimum of 6,000 persons in the U.S. are allergic to aspartame or products containing it.

Dr. N. L. Novick (1985) documented the following unusual reaction to aspartame in the form of inflammation affecting fat tissue (referred to as granulomatous panniculitis).

A 22-year-old woman developed leg lesions over a two-month period that were consistent with erythema nodosum on microscopic examination. She had switched to an aspartame-sweetened diet soda ten weeks previously. These nodules disappeared after avoiding it for one month. Within 10 days after resuming the beverage, many more comparable nodules appeared . . . and again abated after cessation. Comparable lesions reappeared on her legs within 10 days after challenge with pure aspartame.

OTHER ASPARTAME-ASSOCIATED ALLERGIES

Respiratory Tract

Clinically significant allergic reactions to aspartame products have involved the nose, sinuses and bronchial tubes.

A 48-year-old clerk with a history of hay fever and asthma began using various aspartame-sweetened soft drinks and four packets of an aspartame tabletop sweetener daily. After 11 months, her asthma "worsened to the point of being life-threatening—that is, many emergency room visits, and two hospitalizations between October 1985 and July 1986." Concomitant complaints included decreased vision in both eyes, severe dizziness, palpitations, attacks of rapid heart action, and her first grand mal seizure.

These problems disappeared when she avoided all aspartame products. She wrote, "Since stopping the consumption of aspartame in July 1986, I have not had a single asthma attack."

Mouth and Throat

The lips, mouth, tongue, throat and larynx became swollen in some aspartame reactors. Such swelling may be life-threatening, especially in children.

A 60-year-old diabetic woman predictably developed severe swelling of her mouth with itching after eating "dietetic" pudding or aspartame candy.

A 62-year-old woman began to experience difficulty in swallowing shortly after ingesting an aspartame product. She reported, "I was eating a cereal and did not know that it contained aspartame. My throat became paralyzed and I could not swallow. My daughter asked if I had checked for aspartame. When I did, that's when I realized I was using it." Her daughter *also* suffered "throat paralysis" from aspartame.

RELATED PUBLIC HEALTH AND LEGAL CONSIDERATIONS

Aspartame products may be increasing "the allergic load" imposed by numerous natural and man-made substances. (This subject has been designated by some as "environmental disease.") The likelihood that it will take decades to unravel the interrelated biological impact of the more than 1,000 *new* chemicals introduced each year is sobering.

Similarly, aspartame products might play a cumulative role in conjunction with other substances that already have sensitized our population. These include multiple immunizations, exposure to an increasing number of potentially antigenic substances (including widely used drugs), and even vasectomy (Roberts 1979).

The Superior Court of New Jersey underscored the importance of "certain individuals" who "may be allergic or hypersensitive to the product" in *D'Arienzo v. Clairol, Inc.,* (1973). It stated: "Though the nature of the allergic response may be unknown to the average consumer, this is hardly esoteric knowledge within the community that manufacturers these products."

CHAPTER NINETEEN

Gastrointestinal Problems

*That man can interrogate as well as observe
nature, was a lesson slowly learned in his
evolution.*

— S I R   W I L L I A M   O S L E R

SEVERE GASTROINTESTINAL REACTIONS TO aspartame-containing products can develop anywhere from the mouth to the rectum. Their contributory role now must be considered when gastrointestinal complaints do not respond to conventional treatment.

The diagnosis of these reactions may be difficult, especially if stomach or bowel disorders previously existed. For example, they occurred among patients with histories of hiatus hernia, peptic ulcer, "irritable bowel" ("spastic bowel," "nervous colitis"), inflammatory bowel disease (ulcerative colitis; regional enteritis or Crohn's disease), pancreatitis, diverticulitis, hemorrhoids, and anal fissure. In such cases, the key to the diagnosis was the relief of symptoms after stopping aspartame products, and their recurrence on rechallenge— usually as deduced by a perceptive patient.

A 38-year-old woman developed severe abdominal pain, nausea, diarrhea with bloody stools, and marked abdominal bloat over the 18-month period during which she drank 3 to 4 cans of diet cola daily. She described her ordeal.

"I was sick for over a year and a half as a direct result of ingesting aspartame. My reaction was, I believed, purely intestinal in na-

ture . . . continual diarrhea and severe stomach upsets. A gastro-
enterologist put me through many upper and lower GI tests,
several absorption-rate tests, and one particularly nasty "string"
test for parasites, but could not determine what was causing my
nausea and diarrhea. He never questioned what I was ingesting.

"Becoming desperate, I started reading everything that re-
motely described my symptoms. With aspartame and food allergies both receiving media attention, I realized my symptoms
appeared (a) right after we moved into our new house, and (b)
approximately the same time aspartame was introduced into cola
products. I immediately stopped consuming any soft drinks.
After ten days of uncomfortable withdrawal symptoms, I felt
like a new person."

The similarity of aspartame-related reactions involving the stomach and bowel among members of affected families, mentioned in
Chapter 8, was impressive.

The FDA has received comparable aspartame-related gastrointestinal complaints from consumers (Tollefson 1987). They included
254 persons with abdominal pain and cramps, and 178 with diarrhea.

MOUTH, TONGUE AND THROAT REACTIONS

Oral reactions to aspartame products were mentioned under allergies (Chapter 18). In this series, they included swelling of the tongue
and lips (29 patients), and difficulty in swallowing (28 patients). The
latter was erroneously attributed to hysteria in several persons.

Patients also described altered taste, dryness, and other severe
mouth sensations that persisted in spite of prior dental, medical,
allergic and neurological consultation. Persons with these problems
warrant a diagnostic trial period of abstinence from aspartame-containing products.

A 72-year-old woman developed severe reactions of the lips and
tongue. They persisted after avoiding lipstick, at first thought to
be the offender. Although known to be an aspartame reactor, she
did not realize that the ginger ale being ingested had aspartame.
The lip and tongue problem subsided within one week after
avoiding this beverage.

Dramatic throat reactions occurred in both a mother and daughter. The former wrote:

"I was drinking aspartame hot chocolate two or three times a day. One evening while taking a walk, I was unable to swallow. It was as if my throat was paralyzed. I didn't mention it to my daughter because I didn't want to frighten her.

"Several days later, she related to me that she had been drinking a chocolate mix with aspartame every night, and that she couldn't swallow. Needless to say, I was much relieved to finally discover what was causing my throat to be paralyzed.

"We both stopped using the sweetener for a while, and then tried it once more with the same result—inability to swallow. I'm sure the aspartame caused it because we have not had the problem since we stopped using it."

SWELLING OF THE SALIVARY GLANDS

Several aspartame reactors gave histories that were highly suggestive of salivary gland swelling.

A 28-year-old woman developed swelling of the "glands located just under the jaw" on *three* different occasions after ingesting aspartame beverages, puddings, gelatins and gum. Other aspartame-related complaints included swelling of the lips and tongue, abdominal pain, severe nausea, diarrhea, ringing in the ears, dizziness, insomnia, depression, marked irritability, a personality change, attacks of rapid heart action, frequent urination during the day, less frequent periods, thinning of the hair and joint pains. All symptoms abated after stopping aspartame products.

A woman described a similar reaction to an aspartame-containing beverage.

"I had an allergic reaction to a diet cola about a year ago. I drank a glass. About 15 minutes later, my saliva glands swelled. I had knots in my neck as large as golf balls. I went to the doctor. He said it was an allergic reaction to something I had eaten or drank. He gave me an injection of adrenalin and an allergy injection."

NAUSEA AND/OR VOMITING

Severe nausea was prominent in 79 aspartame reactors (see Case Reports in this and other chapters). Some experienced it "almost immediately" after eating or drinking products containing aspartame, with or without vomiting.

STOMACH AND BOWEL IRRITATION; ABDOMINAL PAIN

Seventy individuals developed severe pain in the abdomen while consuming aspartame products. It presumably represented some form of irritation of the stomach, small bowel, pancreas (see below), or combinations thereof.

Several patients suffered severe cramps or "gas," gastric bleeding, and intestinal bleeding (see below) after ingesting such products. The irritation was attributable to a popular aspartame-sweetened effervescent "natural-fiber laxative" taken for constipation and the "irritable bowel syndrome."

It is ironic that aspartame was accidentally discovered to be sweet in 1965 by a chemist who researched its use as a drug for treating peptic ulcer. (He happened to lick his finger to pick up a piece of paper after the heated chemical had run over a flask, and noted the sweetness.)

Representative Case Reports

A man experienced recurrent attacks of severe upper abdominal pain for which he was hospitalized. Subsequent studies of the stomach, duodenum and pancreas proved normal. Each of these episodes occurred within one hour following the ingestion of an aspartame product. There was no recurrence when he avoided it.

★ ★ ★

A 52-year-old woman complained of uncontrollable "gas and letting out air" after using aspartame. She wrote, "It is embarrassing to let gas out when you are shopping or talking to someone." This symptom and the concomitant headache and abdominal bloat recurred every time she challenged herself with an aspartame product.

PANCREATITIS

The upper abdominal pain suffered by several patients shortly after ingesting an aspartame product was possibly due to inflammation of the pancreas. None dared risk rechallenge for fear of provoking another excruciating attack. The following observations are pertinent.

- Significant accumulation (uptake) of radioactively-labeled aspartame by the rat pancreas has been demonstrated within 30 minutes after its administration (Matsuzawa 1984).
- Phenylalanine causes a significant increase in pancreatic enzyme output (Go 1970).
- The administration of an amino acid solution containing phenylalanine is being used to stimulate chymotrypsin production by the pancreas in testing drugs and other substances for their possible benefit in severe chronic pancreatitis (Slaff 1984).

Representative Case Report

A 54-year-old registered nurse drank up to ten glasses of an aspartame-sweetened soft drink and four glasses of aspartame hot chocolate daily. She suffered severe upper abdominal pain, nausea, bloody diarrhea and abdominal bloat. Other complaints included recurrent attacks of hypoglycemia, visual difficulty, headache, dizziness, insomnia, confusion, memory loss, tremors, hyperactivity, tingling of the limbs, intense depression (with suicidal thoughts), extreme irritability, personality changes, palpitations, thinning of the hair, and joint pains. Attributing her "severe attacks of hypoglycemia and pancreatitis, which required frequent hospitalization" to aspartame products, she wrote

"I feel that the pancreatitis was a direct result of aspartame use. I have never used alcohol or drugs of any kind. The doctor cannot find any explanation for it, other than the use of aspartame . . . The pancreatic pain began, along with the mental symptoms, following a vacation trip on which I consumed much more aspartame than while working. It was after consuming a large glass of aspartame on an empty stomach that I passed out and was rushed to intensive care."

DIARRHEA

Seventy reactors suffered diarrhea while using aspartame products. The stools were *grossly bloody* in 12.

A 35-year-old woman developed diarrhea after drinking several cans of a diet cola. Her 54-year-old mother also experienced diarrhea from this product. She offered these perceptive comments:

"Tolerance levels seem to vary. Only one or two cans of the soft drink affect my mother immediately. For me, it took more. If I had one glassful during the day—no problem. But after several throughout the day or for two days, the problem will begin."

An 18-year-old college student suffered abdominal pain, nausea, and bloody diarrhea after consuming one liter of diet cola daily. Other complaints included dizziness, headache, intense drowsiness, palpitations, and marked thirst. All subsided within ten days after avoiding aspartame beverages. The diarrhea and abdominal pain promptly returned on the one occasion she challenged herself with an aspartame drink.

RELATED PUBLIC HEALTH CONSIDERATIONS

Many aspartame reactors in this category expressed outrage over the following issues.

- The failure of multiple physicians to have considered aspartame products as a possible cause for their distress.
- The risks and costs of the gastrointestinal x-rays, CT scans and endoscopic studies to which they had been subjected.
- The significant amount of diagnostic radiation received, particularly by young persons suspected of having peptic ulcer, ulcerative colitis and Crohn's disease before the diagnosis of reactions to aspartame products.

CHAPTER TWENTY

Heart and Chest Problems

*But the interesting thing—or better, the tragedy—
is that the most dangerous items provoke no fear at
all. We decide* caveat emptor *when, but only
when, it is convenient to do so.*
— *DR. JOHN B. THOMISON*
*(1983)**

T HE FREQUENCY AND REPETITIVE nature of certain heart and chest complaints in persons having adverse reactions to aspartame-containing products were surprising. They included attacks of rapid heart action, other types of "palpitation," "shortness of breath," atypical chest pain, and unexplained hypertension (high blood pressure). Again, the crucial clues were improvement of these features after stopping such products, and their prompt recurrence following rechallenge.

ABNORMAL HEART RHYTHM

Eighty-eight aspartame reactors in this series experienced a disturbing change of their heartbeat after consuming aspartame products. It ranged from episodic "fluttering" to rapid heart action (tachycardia). Of the 397 persons who completed the questionnaire,

* © 1983 *Southern Medical Journal.* Reprinted by permission from the *Southern Medical Journal* and John B. Thomison, M.D.

24 had to undergo heart monitoring (Holter) studies for these complaints and concomitant weakness or faint feelings.

Representative Case Reports

An 82-year-old woman had enjoyed good health until using an aspartame-sweetened soft drink. She then "began having cardiac symptoms—palpitations, flutterings, premature beats, and feelings that the heart was getting up and turning over. My doctor exhausted all the tests he could think of doing. Nothing seemed amiss. I was plagued with those symptoms around the clock, regardless of activity."

The patient herself then put "2 and 2 together, and came up with aspartame." Her symptoms disappeared two days after stopping this product, and did not recur. She asserted, "To me, that sounds as if aspartame was the culprit."

* * *

A woman developed an "abnormal heart beat" accompanied by difficulty in breathing, a profuse sweat and uncontrollable tremors *the first day* she used an aspartame tabletop sweetener. These complaints promptly disappeared after she avoided aspartame products based on "my own detective work". They recurred after she ingested one not suspected of containing aspartame.

SHORTNESS OF BREATH

Fifty-four aspartame reactors—most without known heart or lung disease—specifically emphasized "shortness of breath" (dyspnea). This symptom *promptly* and *predictably* disappeared after cessation of aspartame. A number of these patients had undergone extensive "negative" workups for suspected coronary heart disease, emphysema, atypical asthma, and clots to the lung (pulmonary embolism.)

The cause and significance of this complaint are not clear. However, aspartame probably can affect certain neurotransmitters that influence the brain's respiratory center.

Sleep apnea refers to periods of relatively prolonged absence of breathing during sleep. It dramatically vanished in one male patient (whom I had planned to refer to a special sleep apnea center) after he stopped using an aspartame product. It has not recurred for over a year.

ATYPICAL CHEST PAIN

Forty-four aspartame reactors anguished over unexplained chest pain and its implications. The quality of the pain can be inferred from the fact that 33 underwent exercise stress testing for coronary heart disease. (The results in most cases were normal.)

It is obvious that physicians must first consider a number of diagnostic possibilities when patients present with atypical chest pain. They include heart disorders, pleurisy due to various causes, pulmonary embolism, tumors, chest wall tenderness, esophageal problems, and vertebral collapse in postmenopausal women with osteoporosis.

When noninvasive studies prove normal in such individuals who consume large amounts of aspartame products, a trial of abstinence might obviate invasive studies of the coronary arteries (coronary arteriography.) It is pertinent to point out that chest pain *not* attributable to coronary artery disease now constitutes a major problem as an increasing number of patients are being admitted to coronary care units. In fact, persons in this category account for up to half the admissions to such units (Editorial, *The Lancet* 1:959–960, 1987).

RECENT HYPERTENSION

Thirty-four aspartame reactors without previously-known hypertension were found to have an elevated blood pressure—systolic, diastolic, or both. (It may have contributed to the concomitant headache in some.) In the majority, the blood pressure promptly returned to normal after stopping aspartame products. Similarly, Shabin and Albert (1988) indicated that patients with hypertension appear more prone to the adverse effects of aspartame.

A 23-year-old male in apparent good health tried to enlist in the Armed Forces, but was turned down because of an elevated blood pressure. His physician found a blood pressure of 212/110 and an elevated pulse rate for which no cause could be determined.

The patient then wondered if the aspartame-containing soft drinks he had been consuming were a factor. After stopping these beverages, his blood pressure and pulse repeatedly were found to be normal—namely, a blood pressure of 123/78, and pulse of 78. At that point, his physician requested that he bring along a can of an aspartame-containing cola to the office. Within *two minutes* of taking *one sip* of the beverage, his pulse increased to 120, and the blood pressure to 140/98.

The physician then called me to verify my encounter of other instances of aspartame-associated rapid heart action and elevated blood pressure. By coincidence, I was in the process of returning the galley proofs of my manuscript to the *Journal of Applied Nutrition* that listed dozens of persons with palpitations, tachycardia, and recent hypertension ascribed at least in part to aspartame-containing products (Roberts 1988e).

A number of *mechanisms* could be involved in inducing high blood pressure in aspartame reactors. A few are mentioned briefly.

- Aspartame may elevate blood pressure by increased norepinephrine, epinephrine and dopamine within the nervous system. All are derived from phenylalanine, its major component.
- Some patients who consume considerable aspartame products seem to develop a craving for salt (sodium chloride) . . . as well as for sugar, sweets and caffeine (colas). In conjunction with their greater thirst (Chapter 21), an increased intake of sodium could contribute to a rise in blood pressure (and fluid retention.)
- Previously effective antihypertensive medication (as methyldopa or Aldomet®) became ineffective in several patients, possibly because of apparent drug interactions with aspartame (Chapter 22).

CHAPTER TWENTY-ONE

Other Reactions

*There is a sickness which puts some of us in
distemper, but I cannot name the disease; and it is
caught of you that yet are well.*
                    — *S H A K E S P E A R E*
              (The Winter's Tale, *Act 1, Scene 2*)

I HAVE DESCRIBED REACTIONS to aspartame-containing
products as the new "great mimic" in medical practice (Roberts
1988e) because of their wide clinical diversity. This chapter briefly
reviews some reactions not discussed previously.

The criteria for these aspartame-associated disorders are the same
as mentioned previously—namely, improvement shortly after avoiding such products, and prompt recurrence or aggravation on
rechallenge.

MENSTRUAL AND PREMENSTRUAL CHANGES

One out of ten female aspartame reactors in the appropriate age
range developed significant menstrual changes that could not be otherwise explained. They probably reflect dysfunction of major neurotransmitters within the hypothalamus, pituitary gland and brain
(Chapter 5).

Excessive Menstrual Bleeding

Twenty-three women experienced excessive menstrual bleeding, more frequent periods, or both, while consuming aspartame products. As a result, some were subjected to uterine dilatation and curettage ("D & C")... or even a hysterectomy. Such intervention occurred more often among those over 35. The majority resumed having normal periods after avoiding aspartame products.

A 31-year-old woman developed more frequent menses while using two cans of an aspartame cola, two packets of an aspartame tabletop sweetener, and two glasses of presweetened iced tea daily. She also complained of associated depression and anxiety. The role of aspartame seemed convincing. She stated, "My menstrual cycle occurred approximately every 14 to 21 days, rather than 28 days. Since I have stopped using ANY products containing aspartame, it has never failed to occur exactly every 28 days." The same sequence recurred within two weeks when she rechallenged herself.

A 49-year-old woman noted excessive and frequent menses, as well as severe hives, several weeks after drinking considerable amounts of aspartame-sweetened sodas and iced tea in the summer. These complaints disappeared during the fall, only to recur the following summer when she again increased her intake of aspartame drinks. A uterine biopsy was reported as normal. Her periods normalized shortly after avoiding all aspartame products.

A 17-year-old student recorded "severe menstrual bleeding" as her only reaction to aspartame beverages. She noted, "If I accidentally drink aspartame, I start severe bleeding." Her mother and father also experienced reactions to aspartame products.

Decreased or Absent Menses

Twenty-two women noted a dramatic diminution or loss of periods while consuming aspartame products. Since their menses generally normalized following abstinence, the suspected diagnosis of "early menopause" was clearly erroneous.

Excessive prolactin production by the pituitary gland is a significant cause of menstrual changes or loss in women. Carlson (1989) clearly demonstrated that phenylalanine and tyrosine are the most potent stimulators of prolactin secretion by single amino acids, accounting for most such activity by a mixed meal.

A 45-year-old interior designer had no menstrual period after beginning to take "diet food meals" and four cans of an aspartame cola. Her two teenage daughters also stopped menstruating when they used these products. The menses normalized *in all three women* within one month after avoiding aspartame.

A 30-year-old woman developed decreased vision and pain in both eyes, severe headache, dizziness, and intense thirst when she drank aspartame-sweetened tea daily. The change in her menses proved even more disconcerting. She wrote

"My menstrual cycle was disrupted and my gynecologist could find no reason for this. I stopped using the iced-tea product and my cycle became normal. The severe headaches ceased. I then tried the product again, and these symptoms recurred."

Increased Premenstrual Complaints

Aspartame reactors repeatedly volunteered that their nervous tension, swelling and other complaints referable to the premenstrual syndrome (PMS) were more severe while using products with this additive.

A 33-year-old woman experienced more frequent menses, marked breast tenderness, and prolonged PMS symptoms while consuming diet cola. These features disappeared after abstinence, and promptly recurred following retrial.

INCREASED THIRST

One out of ten aspartame reactors experienced intense thirst and severe dryness of the mouth. These symptoms occurred in the absence of excessive sweating or hot weather. An engineer commented that aspartame "seems to create thirst, not quench it."

The problem of aspartame-associated "dry mouth" (xerostomia) tends to be most severe among older persons. Furthermore, aspartame appears to intensify the reduction of saliva associated with aging (especially in the Sjögren syndrome, wherein tears and saliva are markedly reduced), and the side effects of many drugs. Severe complications of dry mouth may ensue. These include (1) inflammation of the tongue, lips and mouth, (2) difficulty with speech, taste, mastication and swallowing, and (3) superimposed yeast infection.

A vicious cycle can be set into motion when persons drink even larger quantities of aspartame-containing beverages to quench their thirst and to moisten their lips or mouth. Since taste impairment does not necessarily accompany the chronic absence of saliva (Weiffenbach 1987), such individuals may develop a preference for the taste of aspartame in satisfying their thirst . . . thereby perpetuating the dry mouth syndrome.

> A 34-year-old woman described constant and severe thirst as her main problem with aspartame products. She also experienced decreased hearing in both ears, phobias, and otherwise-unexplained attacks of shortness of breath. All of these symptoms disappeared after abstaining from aspartame products for one month. They recurred *within one day* on each of the *four* occasions she rechallenged herself with such a product.

FLUID RETENTION

The retention of fluid became a major problem for some aspartame reactors, especially women subject to the premenstrual syndrome (PMS). Its clinical manifestations included marked "bloat" (in 57 patients), and swelling of the lower limbs (in 20). One female likened the recurrent "moon face" during aspartame-induced attacks to that of persons taking cortisone-like drugs. The majority suffered concomitant headache, depression and severe tension. These complaints usually regressed shortly after avoiding aspartame products.

> A 46-year-old woman predictably experienced abdominal swelling and discomfort *within one day* after drinking aspartame beverages during *four* rechallenges. ("When I drink two regular glasses of drinks with aspartame, I feel bloated and have diarrhea the next day.")

> A female executive suffered severe abdominal bloat after drinking aspartame sodas. She stated

"The worst thing I noticed was that I was bloated all of the time, especially at night when I awoke and would have to make six trips to the bathroom. I went to the doctors several times to see if I had a bladder infection, but they did not diagnose anything.

"Why do I blame aspartame? Because I only really take diet drinks in the summer, and the symptoms disappear when I stop drinking them. I have not touched it for about three weeks and I feel great. The bloating is gone, and I have lost three pounds without starving myself."

URINARY COMPLAINTS

Increased frequency of urination (day, night, or both) was a prominent complaint in 69 aspartame reactors—both male and female. The excessive consumption of diet drinks containing caffeine (which acts as a diuretic) contributed to the increased volume of urine. In addition, amino acids (as well as a high-protein diet) tend to increase blood flow to and within the kidneys (Klahr 1988).

Other patients predictably suffered severe burning on urinating after rechallenge with aspartame products. Such irritation of the urinary bladder or urethra may be attributable to aspartame or its breakdown products, particularly since the D-forms of aspartic acid and phenylalanine tend to be poorly handled by the body, and at least half are excreted in a highly acid urine (Man 1987).

Some men in this series had undergone prostate surgery for these complaints. Others planning to have it cancelled the operation when their symptoms subsided after avoiding aspartame products.

A 40-year-old woman had frequency of urination during the day, and voided four times at night, while consuming up to ten packets of an aspartame tabletop sweetener and several aspartame soft drinks daily. On one occasion, she became incontinent while sleeping. Marked thirst, "a general feeling of ill health," and severe depression were concomitant complaints. All persisted in spite of consultation with three physicians. The patient herself deduced that these symptoms were related to aspartame products when they (1) regressed shortly after avoiding them, and (2) predictably recurred within "several hours to overnight" during *three* rechallenges with as little as one diet soda or a packet of tabletop sweetener. She stated

"If I hadn't stopped using aspartame, my next referral was to a urologist for a urethral dilation as suggested by my gynecologist. Fortunately, I noticed a severe increase in symptoms after two cans of diet soda, and began to research it on my own."

RHEUMATIC COMPLAINTS

My experience suggests that unexplained rheumatic discomforts in persons using aspartame products warrant a trial of avoidance before embarking upon expensive diagnostic studies and potent drugs.

Fifty-eight aspartame reactors in this series suffered joint pains (arthralgia) that improved or disappeared after abstaining from this additive. The concomitant use of considerable aspirin intensified other aspartame-related complaints—particularly dizziness, ringing of the ears, and hearing loss.

A 45-year-old telephone technician developed "severe joint irritation" and "less than 1/4 my normal strength" after consuming two packets of an aspartame tabletop sweetener. His disability regressed just one day after stopping this product. On the *three* occasions he retested himself, the joint symptoms recurred within one day.

A 62-year-old male stated that *"all* my joints ached *all* the time" while using up to eight packets of an aspartame tabletop sweetener in coffee, one glass of an aspartame hot chocolate, and two servings of aspartame puddings or gelatin daily for four months. Other complaints included loss of vision in one eye, severe headache, drowsiness, sensitivity to noise in both ears, unexplained facial pains, tingling of the limbs, extreme irritability, unexplained chest and abdominal pains, and a gain of 30 pounds. All symptoms abated after stopping aspartame products—only to recur within eight hours during *two* rechallenges.

A 57-year-old engineer complained of "severe arthritic-like pains in most major joints" during 1984. He had been drinking 10 to 12 cans of an aspartame cola since 1983. A rheumatologist diagnosed his problem as "an autoimmune response, most likely from a bacterial infection," and prescribed Feldene® (an anti-inflammatory drug) for two years. Within two weeks after stopping the aspartame beverage, he felt markedly improved.

INCREASED SUSCEPTIBILITY TO INFECTION

Some aspartame reactors developed recurrent infections that could not be directly attributed to any underlying disorder, such as diabetes. Admittedly, the loss of diabetic control (Chapter 25) or exaggeration of hypoglycemia (Chapter 26) associated with aspartame use may be conducive to infection in various organs, such as the prostate.

A 58-year-old man experienced repeated attacks of prostatitis over one year, requiring maintenance trimethoprim and sulfomethoxazole (Bactrim®). A urologist had been unable to find any contributory cause. The possible role of aspartame was raised on the basis of recent unexplained headache, dizziness, visual problems, confusion with profound memory loss, "dry eyes," drowsiness, and severe irritability. The patient had been drinking two cans of an aspartame cola and three glasses of aspartame-sweetened iced tea daily. His complaints improved within one week of abstinence. They did not recur over the subsequent seven months of observation.

Related Considerations

Several related issues warrant further study. The possibility that an effect on the immune system induced by aspartame products might be contributory to recent near-epidemics of certain infections, or suspected infections, is intriguing. In experimental studies involving the diketopiperazine metabolite of aspartame (Chapter 5), three outbreaks of infectious disease required treatment with penicillin.

- Some persons with diagnosed or self-diagnosed chronic infectious mononucleosis ("yuppie flu") due to the Epstein-Barr virus—based on chronic fatigue, sweats, weight loss, headache and joint discomfort—may actually be aspartame reactors.
- Acute rheumatic fever ("the red plague") has reemerged since 1984. Serious outbreaks have occurred among affluent families . . . contrasting with the environment of poverty in which

it developed earlier this century. Moreover, two-thirds of such children did not have an obvious Streptococcus infection preceding the attack even though this bacterial organism usually could be cultured from siblings.

- In my opinion, a possible deleterious influence of excessive aspartame consumption in the acquired immune deficiency syndrome (AIDS) requires investigation. Many patients with HIV infection use aspartame products instead of sugar in an attempt to decrease the risk of fungal and other superimposed infections.

CHAPTER TWENTY-TWO

Drug Interactions

More is missed by not looking than by not knowing.

— *THOMAS MOORE*

OBSERVATIONS BY MYSELF AND others suggest that clinically significant interactions may occur between aspartame-containing products and commonly used drugs. The drugs include coumarin (Coumadin®), phenytoin (Dilantin®), antidepressants, propranolol (Inderal®), methyldopa (Aldomet®), insulin (Chapter 25), and lidocaine (Xylocaine®).

Such an influence may be manifested as either reduced or excessive drug activity. Mention was made earlier of impaired levodopa action in treating Parkinsonism (Chapter 12) by amino acids and protein. A few of the many *possible mechanisms* of drug interactions are listed.

- Alteration of the blood proteins to which drugs attach.
- Altered receptors on the membranes of cell walls.
- Interference with drug transport across the blood-brain barrier.
- Existing metabolic disturbances and impaired kidney function in the elderly that enhance their vulnerability to chemicals and drugs (Weber 1986).
- An interaction between the methanol component of aspartame and various ethanol-related compounds (Chapter 6). The latter include the hypoglycemic sulfonylureas (used for treating diabetes), metronidazole (an antibacterial drug), allo-

purinol (used in the treatment of gout), and disulfiram (Antabuse®) taken as maintenance prophylaxis by alcoholic patients (Chapter 27).

COUMARIN (COUMADIN)

Coumarin is widely used as an anticoagulant ("blood thinner") in the treatment or prevention of diseases resulting from thrombosis (clot formation) or embolism (the migration of a thrombus). These conditions include coronary heart disease, strokes due to embolism from the carotid artery or from the heart in patients with irregular heart action (atrial fibrillation), embolism to arteries in the lower extremities, and pulmonary embolism secondary to thrombophlebitis in the lower limbs and pelvis.

Treatment with coumarin represents a major therapeutic commitment because it necessitates careful monitoring to keep the blood prothrombin time within safe boundaries. A patient may bleed if the prothrombin time becomes too prolonged. On the other hand, loss of the anticoagulant effect invites recurrence of the underlying clotting disorder.

Interference by aspartame products on the prothrombin time was suspected in several patients who had been on coumarin for prolonged periods. Their prothrombin times unexpectedly declined (meaning a loss of anticoagulant effect) after consuming aspartame, and was then followed by the recurrence of thrombophlebitis or angina pectoris.

A 72-year-old woman received coumarin for angina pectoris and thrombophlebitis. Her prothrombin time was under good control, generally ranging between 19.5–22 seconds (control values 11–12 seconds). The patient unexpectedly developed unstable angina pectoris and rapid heart action (tachycardia). Hospitalization was required. At the time, little significance was paid to the slight decline of her prothrombin time (17 seconds). It decreased even more—to 15 seconds—at discharge, notwithstanding an increase in coumarin dosage.

A possible influence of aspartame on the prothrombin time was suggested by concomitant complaints. They consisted of severe headache, "burning and swelling" of the lips and tongue, visual problems (in spite of recent cataract surgery and prescription glasses), and unexplained nausea. The "diet" soda she was drinking contained aspartame. Her prothrombin time rose to 22 seconds one week after avoiding it. There was concomitant striking improvement of the fatigue and lip-tongue reactions.

PHENYTOIN (DILANTIN)

Phenytoin is a key drug used in treating epilepsy. Several of the therapeutic problems associated with intake of aspartame products in patients receiving phenytoin were described in Chapter 9. They include the need for increasing the dose of phenytoin, the addition of other anti-epilepsy drugs (each having its own significant side effects), and "rebound" phenytoin toxicity. The latter occurred when higher doses of phenytoin were continued after the patient stopped aspartame products.

A 28-year-old man developed seizures after a head injury while serving in the Marines. They had been controlled with several anti-epileptic drugs, but unexpectedly recurred after drinking large amounts of diet sodas. He had to be taken to an emergency room for phenytoin intoxication shortly after stopping these beverages— presumably representing a rebound phenomenon.

ANTIDEPRESSANT AND OTHER
PSYCHOTROPIC DRUGS

Aspartame-containing products appear to pose a threat to some patients with depressive illness on several accounts.

- They may produce or precipitate severe depression, even among persons having no prior history of depression.
- Aspartame products appear to interfere with the action of major antidepressant medications, such as the tricyclic antidepressants (e.g., imipramine or Tofranil®) and the monoamine oxidase (MAO) inhibitors. The latter include phenelzine (Nardil®), isocarboxazide (Marplan®), and tranylcypromine (Parnate®).
- Aspartame or its components and metabolites may exaggerate the adverse effects of the MAO inhibitors. It is known that hypertensive crises can occur in patients receiving these drugs

after they consume foods and beverages containing tyramine and tryptophan, another amino acid. (This sequence probably represents vasospasm caused by amino acid-derived sympathomimetic substances, notably norepinephrine and tyramine.)

Other physicians have encountered comparable drug reactions. The following observations by Walton (1986) are germane.

A 54-year-old woman had recurrent major depressions that were controlled with a maintenance dose of 150 mg imipramine at bedtime. She then developed a grand mal convulsion and manic behavior. It was ascertained that she had been drinking considerable amounts of aspartame-sweetened iced tea for several weeks prior to the seizure in an attempt to lose weight. The manic behavior subsided within four days after starting lithium and avoiding the aspartame drink. When the depression recurred two months later, imipramine was successfully resumed in the previous dosage. There were no recurrences of severe depression or manic activity over the next 13 months.

The prescription of "anti-anxiety" and other psychotropic drugs for aspartame-related nervousness may compound the problem if a drug interaction occurs.

A 57-year-old engineer developed progressive nervousness, insomnia, memory loss, speech difficulties and severe arthritis pain while consuming 10–12 cans of an aspartame cola and 4–6 packets of an aspartame powdered sweetener daily. He was placed on Xanax® (0.25–0.5 mg three times daily) by a psychiatrist—only to become "more moody and nervous." A shift to small doses of Tofranil® triggered a "near-panic attack." Within two weeks of abstinence from aspartame products, he felt markedly improved in all respects.

PROPRANOLOL (INDERAL)

Benign ("essential") tremor usually can be controlled with small or modest doses of propranolol. Some of my patients and correspondents found that such tremor was intensified by aspartame-containing products, notwithstanding increased doses of propranolol (Chapter 12). The cessation of such products resulted in prompt improvement.

METHYLDOPA (ALDOMET)

Seven of the 397 reactors who completed the survey questionnaire had been on this drug when aspartame-associated complaints began. Reference to aspartame-associated seizures and other disorders among patients receiving methyldopa was made in the two U.S. Senate hearings on aspartame held during 1985. The observation that aspartame administration decreases the entry of methyldopa into rat brain (Maher 1987) is relevant.

CHAPTER TWENTY-THREE

Pregnant Women; Nursing Mothers

By far the most mutagenic agents known to man are chemicals, not radiation. And in this regard, food additives rather than fallout at present levels may present a greater danger.
— *DR. RICHARD CALDECO*
(1961) *(Atomic Energy Commission)*

IN MY OPINION, *PREGNANT women and nursing mothers should avoid aspartame-containing products.* Let me summarize the basis for this assertion.

- An elevation of phenylalanine concentrations during the third trimester of pregnancy can interfere with growth of the fetus brain.
- Phenylalanine levels may be as much as four times higher on the fetal side of the placenta.
- Aspartame-associated weight loss (Chapter 16) by a pregnant woman, or her failure to gain adequate weight, can have disastrous effects on a fetus. (Such concern increases in the case of pregnant teenagers with poor nutritional and other habits.)
- Aspartame reactions consisting of nausea, vomiting and diarrhea (Chapter 19) could result in maternal malnutrition.
- No fetus or infant should be knowingly exposed to methyl alcohol, a poison (Chapter 6).

The Unresolved Issue of Birth Defects

The incidence and nature of birth defects associated with use of aspartame products during pregnancy are not yet known. Concern about a possible thalidomide-like problem has been raised because aspartame has been approved as a safe additive for pregnant women and infants. No extensive trials with aspartame products were done on pregnant women prior to its licensure . . . as required for new drugs.

I have encountered several dramatic instances of severe abnormalities affecting the fetus or the infant of parents who consumed considerable aspartame products at the time of conception, during pregnancy, or both. The parents and grandparents of children born with congenital deformities and other disorders, whose mothers ingested aspartame products while pregnant, have vented their confusion and frustration.

Admittedly, a few encounters in practice do not qualify as a significant epidemiologic study. But neither can they be totally ignored—especially relative to labeling precautions for consumers (Chapters 30–32).

> A young couple ingested large amounts of aspartame products prior to and during the wife's first pregnancy. The child was born with several defects, and died three days after birth. I was made aware of this when the husband later consulted me because of aggravated reactive hypoglycemia that could not be explained (Chapter 26). The symptoms promptly abated after avoiding all aspartame preparations; his wife also stopped using them at the time. When seen two years later, the proud father pulled out a picture of a "completely healthy" 9-week old son.

There is a small but increasing body of clinical and experimental information pertaining to congenital malformations and abnormal central nervous function among infants whose mothers took aspartame while pregnant or who had elevated phenylalanine levels.

- My registry contains reports of a group of children with aspartame-induced seizures in the *same* family whose mothers had ingested aspartame products during pregnancy and while nursing.
- Cleft lips and cleft plates are more frequent among persons

having high blood phenylalanine levels (Tocci 1973) and phenylketonuria (Chapter 29).

- Lewis and colleagues (1985) administered aspartame to rats prior to conception. The slightly elevated maternal phenylalanine and tyrosine blood levels in the rats were associated with small brains (microcephaly) and lasting behavioral problems (hyperactivity, learning difficulties) in the offspring.

- Dow-Edwards and Deibler (1989) have noted developmental and neurobehavioral effects of aspartame in experimental models, which correlated with changes in maternal and fetal plasma amino acid levels. In one study where aspartame was given to pregnant guinea pigs, they found the following: (1) blood phenylalanine and tyrosine levels increased two to three times in both the mother and fetus at the three-fourth's stage of pregnancy, while the other large neutral amino acids (leucine, isoleucine and valine) decreased; and (2) the blood phenylalanine and tyrosine levels in the pups were normal on the 20th day after birth, but the other large neutral amino acids were significantly elevated.

Representative Case Report

A 34-year-old woman consumed two cans of diet soda, one glass of another aspartame beverage, one or two packets of an aspartame tabletop sweetener, one glass of presweetened iced tea, and two bowls of aspartame-sweetened cereal daily for two months. Such intake began immediately prior to conception, and continued through the first two months of pregnancy. She developed severe headaches, lightheadedness and foot numbness during this period. Repeat ultrasound studies demonstrated fetal malformations. Abnormalities also were found by amniocentesis. The pregnancy was terminated at the 18th week. An autopsy on the fetus revealed deformed limbs, unusual skin changes, and chromosomal abnormalities consistent with the Turner syndrome.

CHAPTER TWENTY-FOUR

Infants and Children

*Parents have been reporting these findings to
pediatricians for years, and most of those reports
have been dismissed as anecdotal. The key here is
that anecdotal evidence is acceptable when it
agrees, but dismissable when it conflicts with
prevailing ideology. These patients apparently were
quite accurate in their accounts to physicians, even
in the face of contrary prevalent medical
opinion . . . But the history of the environmental
movement has provided an alarmingly regular
parade of cases in which the experience of
individuals with environmental contaminants was
dismissed as unscientific by the established
authorities, only to be later confirmed scientifically.*
— K E N   A N D   J A N   N O L L E Y
(1987)★

RESERVATIONS CONCERNING THE POTENTIAL harm
to infants and young children from aspartame-containing products
reinforce those presented in the preceding chapter. They recall the
effect on Alice of drinking the contents of a little bottle labeled with
the words "Drink Me" beautifully printed in large letters.

In my opinion, parents should minimize—or prohibit—the consumption of

★© 1987 *The Journal of Pesticide Reform.* Reproduced with permission.

aspartame products by their infants and young children until the matter of potential harm is convincingly clarified.

- Nursing mothers, I believe, should not consume aspartame products while breast-feeding. The case of a suckling infant who developed seizures as its mother drank an aspartame beverage offers a dramatic illustration.
- Approximately five servings of an aspartame pudding by a 50-pound child, or a comparable amount of other aspartame products, equals the maximum advisable daily intake (ADI) of 50 mg aspartame/kg body weight.
- It may take a generation or longer to ascertain the full extent of developmental abnormalities, and of intellectual or behavioral sequelae, attributable to aspartame products taken during pregnancy and in childhood. (A comparable situation formerly existed in the case of children whose mothers drank alcohol during pregnancy.) One aspartame reactor wrote me, "So many children are raised drinking and consuming products containing aspartame that we may never know the impact on their lives."

Others have expressed similar concern.

- Senator Howard Metzenbaum asserted in a March 3, 1986 news release: "We cannot use America's children as guinea pigs to determine 'safe' level of aspartame consumption."
- Dr. Louis Elsas of Emory University testified at a U.S. Senate hearing on August 1, 1985: ". . . there's no reason why a child less than six months old should be taking aspartame."

ADVERSE EFFECTS

Aspartame-associated convulsions (Chapter 9), headaches (Chapter 10), rashes (Chapter 18), asthma (Chapter 18), anorexia (Chapter 16), gastrointestinal problems (Chapter 19), and severe acidosis (Chapter 6) were considered in earlier sections.

Additional Representative Case Reports

An 8-year-old girl experienced severe headaches daily, requiring repeated doses of aspirin. As arrangements were being made for her to be seen in medical consultation, the child told her mother that her headaches *predictably* occurred *ten minutes* after chewing an aspartame-containing gum. They promptly ceased when the gum was avoided.

★ ★ ★

A 7-year-old girl developed severe diarrhea shortly after drinking three glasses of an aspartame soft drink mix. It stopped when she abstained from drinking it, but promptly recurred on the *four* times she resumed the drink.

★ ★ ★

A 2-1/2 year-old girl had been consuming up to six cans of an aspartame diet drink and two glasses of a soft drink mix daily when she developed a seizure lasting 45 minutes that affected the right side of her body. She was subjected to two sets of head x-rays, a CT scan and an MRI study of the brain, and two EEGs. All proved normal. It was concluded that the seizure had been related to the aspartame products. Her mother suffered from aspartame-associated headaches.

Abnormal Behavior

The use of aspartame products by children has been associated with irritability, hyperactivity, crying, whining, aggression, and "Jekyll and Hyde" behavior (also see Chapter 13). Others have made similar observations.

The *Congressional-Senate Record* of May 17, 1985 noted the erratic behavior of a 4-year-old boy who drank an aspartame-containing beverage over a three-week period. His behavior normalized within 24 hours after stopping this product . . . only to recur within 30 minutes after the lad was rechallenged with aspartame two weeks later.

The possible contributory role of aspartame products to criminal behavior warrants special study in view of the increasing rate of criminality by young children and teenagers. The behavioral consequences of aspartame reactions could become superimposed upon the many problems confronting today's teenagers—including conflict with parents, difficulties in communication, and the absence of discipline at home.

Cognitive Problems

Aspartame-related deterioration of intelligence and learning was considered in previous chapters. Unfortunately, slippage in school performance during critical periods of brain maturation tends to be subtle before its full magnitude is realized . . . with possible irreversibility. The risks of aspartame-associated retarded growth and mental development are likely to be magnified when hypothyroidism (underactive thyroid function) is not recognized in infancy (Sokoloff 1967) due to the slower metabolism of drugs and chemicals therein.

There is increasing interest in deranged aspartic acid (aspartate) metabolism relative to neurologic disorders affecting infants. The deficiency of a key enzyme (aspartoacylase) can result in severe psychomotor retardation and so-called childhood leukodystrophy of the brain (Hagenfeldt 1987).

Depression and Other Emotional Disorders

At least *ten percent* of teenagers are afflicted with significant depression at some time. Moveover, the suicide rate in this age group continues to climb. It therefore is unwise to ignore the possible contributory role of a pharmacologically-active additive such as aspartame that is being widely consumed by children. As noted above, the clinical expressions of these effects can become superimposed upon the myriad of social problems already confronting many young people.

Adult aspartame reactors have poignantly expressed concern about their perception of a link between suicide in young persons and the consumption of aspartame products.

RELATED CONSIDERATIONS

Physicians' Attitudes

Many pediatricians fail to express an interest in possible aspartame-associated problems. In this vacuum of perceived apathy, some parents have had to resort to their own researches—including the keeping of careful diaries, and the institution of elimination diets. The indifference and/or limitations of pediatricians and other physicians relative to chemically sensitive children are encompassed in the quotation by Ken and Jan Nolley (1987) at the beginning of this chapter.

The issue becomes more complicated when parents are not even aware that many drug preparations intended for use by children contain aspartame. They include "delicious" penicillin and other antibiotic suspensions, chewable acetaminophen tablets for fever, and popular over-the-counter vitamin products.

Exposure to Radiation

A number of children with aspartame reactions were exposed to considerable diagnostic radiation before intolerance to aspartame products was recognized. These studies included x-rays of the stomach, small intestine and large bowel for suspected peptic ulcer, Crohn's disease or ulcerative colitis, and a gamut of radiologic examinations for suspected diseases of the brain and spinal cord . . . including tumors and multiple sclerosis.

"Planned Malnutrition"

Research pertaining to the increased longevity of experimental animals by decreasing their caloric, carbohydrate and protein intake during the neonatal period must not be perverted through projecting these data onto the nutrition of young children. Yet, some sophisticated parents are deliberately underfeeding their children in an attempt to prevent coronary heart disease, dental carries, cancer, and other degenerative disorders.

Also to be considered is that the administration of aspartame-sweetened foods and beverages to young children could invite a life-long preference for sweets rather than "good nutrition." Exposure of the very young to innovative marketing undoubtedly initiates *memes* (Schrage 1988). These are fragments of thought that remain burrowed in the mind, and—like genes—tend to replicate themselves, and mutate from one host to another. In the case of advertisements for aspartame-containing products, memes may be induced by cute tunes, ideas or catch-phrases that virtually parasitize the mind . . . just as with political symbolism and other forms of intended cultural brainwashing.

"Negative Scientific Studies"

The issue of aspartame-associated abnormal behavior, seizures and other problems in children continues to be a subject of debate, largely because corporate-funded double-blind studies report "negative" conclusions.

In my opinion, most of the investigations to date have been improperly designed because the *same* products actually consumed in daily use were *not* given in the study (Roberts 1988e). Instead of commercially-available aspartame beverages that might have been stored for indefinite periods or exposed to a hot environment, and heated aspartame products (e.g., cereal or hot chocolate), the young subjects in these studies received aspartame capsules or freshly-prepared cool drinks. These products are not really comparable. First, blood phenylalanine levels are lower when given in the capsule form. Second, stereometric rearrangements in the aspartame molecule, and other metabolites occur as a result of heat and prolonged storage. Third, some components of food and beverages (especially chocolate and certain aldehydes) might serve as catalysts for these changes.

Consequently, there should be extreme reservation in accepting "negative" conclusions of such studies at face value—especially when the health of infants and children is at stake.

Legal Ramifications

Several appellate rulings pertain to the protection of a fetus against chromosomal and other changes caused by a defective food product or a drug taken before or after conception.

- Ruling that the previous dismissal was incorrect in *Jorgensen v. Meade Johnson Laboratories* (1973), the United States Court of Appeals, Tenth Circuit, stated

   If the view prevailed that tortious conduct occurred prior to conception is not actionable in behalf of an infant ultimately injured by the wrong, then an infant suffering personal injury from a defective food product, manufactured before his conception, would be without remedy. Such reasoning runs counter to the previous principles of recovery which Oklahoma recognizes for those ultimately suffering injuries proximately caused by a defective product or instrumentality manufactured and placed on the market by the defendant . . . We feel that the trial court was incorrect in concluding that recognition of such a cause of action must await legislative action. (p.240)

- A classic instance of injury to a fetus by a substance taken by the mother during pregnancy, but which might not appear in the child for several decades, involves the use of diethylstilbestrol (DES) for the purpose of attempting to prevent a miscarriage. In *Sindell v. Abbott Laboratories* (1980), the plaintiff alleged that she developed genital cancer because her mother had taken this synthetic estrogen for such a purpose . . . an association affirmed by various cancer registries.

CHAPTER TWENTY-FIVE

Diabetes

*If in the field of diabetes all the pathologic
obscurities have not as yet been completely
elucidated, it is because the knowledge of normal
function is still imperfectly understood.*
— *CLAUDE BERNARD*, *physiologist*
(1877)

IN MY MEDICAL PRACTICE as a diabetes specialist, I have
come to the conclusion that *aspartame-containing products may adversely
affect some patients at ALL stages of this disease. This has been evidenced by
(1) the loss of diabetes control, (2) the precipitation of overt clinical diabetes,
(3) more frequent and severe reactions to insulin or oral hypoglycemic drugs, (4)
failure to follow the recommended diet, and (5) the aggravation or simulation of
complications involving the eyes (retinopathy), kidneys (nephropathy), and
peripheral nerves (neuropathy).*

Admittedly, my experience is diametrically opposite to the current
consensus that aspartame products do not adversely influence diabetes (see below). The American Diabetes Association (ADA) continues
to endorse these products enthusiastically as a "free exchange." My
series includes 58 diabetic patients with adverse reactions to aspartame products (Roberts 1988c).

- *Group A*—18 who developed both high blood glucose concentrations (fasting levels of 140 mg per cent or higher) and
  symptoms of diabetes

- *Group B*—23 who were on insulin and diet
- *Group C*—17 who were on oral diabetic drugs and diet

CLINICAL CONSIDERATIONS

The following observations illustrate a few of the problems that may confront diabetic patients while consuming aspartame products.

- The contributory role of such products in the loss of diabetic control was confirmed by *prompt* improvement of blood glucose (sugar) levels after abstinence.
- Aspartame-associated confusion (Chapter 11) contributed to omitting insulin or an oral drug—or taking extra doses.
- Even when the correct amount of insulin was taken, reactions became more frequent and severe while on aspartame products.
- Some "insulin reactions," including convulsions, were most likely aspartame reactions.
- The development of visual symptoms, headache, dizziness, limb pain, and diarrhea or other gastrointestinal complaints had been attributed solely to "diabetic complications" before the contributory role of aspartame products was considered.

REPRESENTATIVE CASE REPORTS

A 46-year-old man with insulin-dependent diabetes had been in good control for three decades until he consumed several aspartame sodas and up to five packets of a tabletop sweetener daily. He summarized his ensuing experience: "My diabetes went haywire, and I had terrible insulin reactions." His diabetes came under control within one week following abstinence from aspartame products.

★ ★ ★

A 48-year-old insulin-dependent diabetic took aspartame-containing products in an attempt to avoid sugar. She noted, "It increased my blood sugar, making it necessary to take (an extra) shot of insulin to counteract it." The associated nausea and extreme irritability precluded her functioning as a teacher.

★ ★ ★

A 67-year-old diabetic woman had been controlled on diet and Tolinase®. Unexplained elevations of her blood glucose then occurred— namely 180 to 247 mg percent before breakfast, and up to 248 mg per cent before supper. A superimposed fungal infection reflected this loss of diabetic control. She had been recently consuming two to three diet colas daily and other aspartame products. Within one week after avoiding them, she felt better. The fasting values normalized (108, 106, 108 and 82 mg percent), and the pre-supper values ranged from 64 to 198 mg percent.

★ ★ ★

A diabetic man complained of severe changes in vision while consuming four liters of aspartame soft drinks daily. An ophthalmologist reassured him that there was no detectable diabetic retinopathy. The patient then chanced to read an article about aspartame-related eye problems, and promptly improved after stopping these beverages.

RELATED CONSIDERATIONS

Physicians' Attitudes

Most physicians continue to accept unequivocal statements issued by the FDA, the AMA, the ADA, and the manufacturer that aspartame is a safe artificial sweetener for diabetics. This attitude has been reinforced by reports concluding there is "no evidence of alteration of diabetic control because of aspartame ingestion" (Horwitz 1983).

Testimony by Dr. F. Xavier Pi-Sunyer (1987), a physician-representative of the ADA, summarizes this issue. He told a U.S. Senate hearing on November 3, 1987 that the ADA reaffirmed the safety of aspartame-containing products. He also stated the following:

- There had been no input from physicians concerning "significant problems with the use of aspartame or, for that matter, that there is any pattern of complaints regarding this product."
- This Senate hearing in itself was generating "unwarranted anxiety among people with diabetes."
- The ADA objected to spending any additional money on further research that "duplicates current knowledge."

I have published (Roberts 1987a, 1988c, d, e)—or attempted to publish—such informational input. It may not be entirely coincidental that the producers of aspartame products have significantly subsidized many projects by diabetes-oriented organizations and investigators. For example, 10 major research grants—most in amounts of $40,000 or more—were given to the Juvenile Diabetes Foundation, according to its 1988 report. A scenario could be conceived wherein a recipient of such largess might read, as an anonymous reviewer for a peer-reviewed journal, a manuscript submitted about aspartame-associated reactions with a trace of self-serving bias.

Pathophysiologic Mechanisms

Reference has been made (Chapter 17) to one of the mechanisms by which aspartame can trigger excessive insulin release—the so-called cephalic phase of insulin release, anticipating the arrival of food.

- Such insulin release is also triggered by the taste of sweet. When insulin production already has been severely strained, this additional stimulation could aggravate "early" diabetes (Roberts 1964, 1965).
- After individuals have used considerable amounts of aspartame products (and perhaps saccharin) for prolonged periods, insulin secretion and release may become altered—as manifested by higher blood insulin and glucose levels . . . especially if they *also* consume more sugar.*

In the absence of sufficient insulin, phenylalanine levels tend to rise. This

---

* A paradoxical rise in the consumption of sugar has occurred over the past decade coincident with the dramatic increased use of artificial sweetners. In point of fact, many diabetics take cake, pie or ice cream with their aspartame-sweetened coffee and tea. According to Department of Agriculture reports, the per capita consumption of sugar was 130 pounds in 1985, compared to 118 pounds in 1975. Consumption of artificial sweetners (including aspartame and saccharin) rose from 6.2 pounds in 1975 to 17 pounds in 1985.

may reflect the role played by insulin in generating the key enzyme phenylalanine hydroxylase (Chapter 5) involved in its metabolism (Tourian 1975).

Phenylalanine (the major component of aspartame) and other amino acids may alter plasma insulin levels and *increase "insulin resistance"* (Felig 1969). Since most adult diabetics are overweight, it is pertinent that Schmid and colleagues (1989) reported a 3.6-fold greater insulin response to an intravenous infusion of amino acids among obese nondiabetic subjects than in lean subjects.

CHAPTER TWENTY-SIX

Patients Prone to Hypoglycemia

*In the last analysis, we see only what we are
ready to see, what we have been taught to see. We
eliminate and ignore everything that is not a part
of our prejudices.*
— *J. M. CHARCOT, M. D.*

REACTIVE HYPOGLYCEMIA ("low blood sugar attacks") is a common disorder, notwithstanding pronouncements by "authorities" to the contrary. I have conservatively estimated that a tendency for hypoglycemia exists among *one out of three* persons in our society (Roberts 1964a, b,d; 1971b).

Hypoglycemic attacks are characterized by recurring severe weakness ("draining of my strength," "sapping of my energy," "late morning slump," "afternoon letdown"), marked hunger, sweats, nervousness, headache, confusion, other mental changes, and a craving for sweets. These episodes generally occur several hours after eating or when the individual has gone too long without food. The attacks have a tendency to become more intense as the day advances, especially during the late afternoon.

I have described elsewhere the many clinical ramifications of reactive hypoglycemia, its intensification at certain times of the day, and its significance as a forerunner of diabetes in a number of publications (Roberts 1964a,b,c; 1965a,b,c; 1966, 1967a,b,c; 1968; 1971b; 1973).

Dozens of patients with previously-diagnosed reactive hypoglycemia experienced an aggravation of their symptoms while using aspartame-containing products. Other aspartame reactors were not aware that they were subject to hypoglycemia until their complaints intensified after consuming aspartame products. Indeed, much of my interest in aspartame reactions initially was stimulated by the failure of many hypoglycemic patients to respond (or continue to respond) to conventional treatment, including a special diet and supportive measures. In some instances, the severity of aspartame reactions caused patients to resume eating sugar.

Neuropsychiatric Manifestations

The brain requires an adequate and continual supply of body sugar (glucose) for proper function. The neurologic, psychiatric and behavioral disorders precipitated by low blood glucose levels can be serious. They include severe dizziness, tremors, double vision, muscle paralysis, confusion, interference with speech, convulsions, depression, and psychopathic behavior. These reactions may alter the attention, perception, coordination and responses by drivers and pilots (Roberts 1971b).

Other potential cerebral effects incurred by severe and unchecked hypoglycemia—especially during the night—deserve emphasis. They are intense sleepiness or narcolepsy (Roberts 1964, 1971a,b), reading disability (Roberts 1969b), migraine (Roberts 1967d), nonthrombotic stroke (that is, not due to a demonstrable blood clot), epilepsy (1964, 1971b), and a state resembling multiple sclerosis (Roberts 1964b, 1966b,c).

A 40-year-old woman with known severe reactive hypoglycemia reported on her reactions to aspartame products.

"All went well until my soft drinks began using aspartame as an additive.

"This summer had been exceptionally hot in the area where I live. I was drinking several cans of diet sodas and diet cola a day until I began to experience side effects that had me feel as I did before my condition was diagnosed and I gave up sugar.

"Deep depression, extreme nervousness, dizziness, diarrhea, and severe headaches were a daily occurrence. I finally began to pay attention to what I was consuming. Upon halting the intake altogether of the diet drinks, I felt much better within a week . . . I can personally say that aspartame is dangerous for me."

RELATED CONSIDERATIONS

Increased Sugar Consumption

It is well known that hypoglycemia often develops after consuming excessive sugar (for example, candy bars). In this circumstance, the sugar load stimulates insulin secretion, causing the elevated blood sugar to fall abruptly. Human volunteers who consumed aspartame were found to increase their consumption of sugar (Blundell and Hill, 1986). This response may account for the unexpected (paradoxical) increase in sugar consumption by diabetics mentioned in the previous chapter.

Mechanisms of Hypoglycemia

Aspartame can precipitate or aggravate hypoglycemia in several ways.

- It is known that phenylalanine and other amino acids stimulate excessive release of insulin (Floyd 1966, 1970; Reitano 1978, Schmid 1989). The insulin, in turn, could provoke a hypoglycemic attack.
- The tendency of weight-conscious persons to restrict calories and to exercise excessively, in conjunction with using aspartame products, invites a further drastic decline in the blood sugar (glucose) concentration.
- There are important relationships between the blood sugar level, insulin concentrations and aspartate (aspartic acid). Insulin-induced hypoglycemia caused a sharp increase of brain aspartate concentrations in experimental studies, whereas the glutamate and glutamine levels declined (Chapman 1987).

CHAPTER TWENTY-SEVEN

Alcoholism

It is of use from time to time to take stock, so to speak, of our knowledge of a particular disease, to see exactly where we stand in regard to it, to inquire what conclusions the accumulated facts seem to point to, and to ascertain in what direction we may look for fruitful investigations in the future.

— SIR WILLIAM OSLER

I HAVE BEEN IMPRESSED by the intolerance to aspartame-containing products of individuals with alcoholism—whether a previous or current problem. The implications of this phenomenon are evident when the magnitude of aspartame consumption is coupled with the fact that about six percent of adults in our society actively suffer from alcoholism.

The following observations and considerations are pertinent.

- Some non-alcoholic patients experienced severe reactions (including convulsions) to aspartame products after also drinking alcohol . . . generally in small amounts.
- Several persons stressed that their aspartame-related complaints intensified after drinking alcohol mixed with an aspartame beverage.
- Patients with a history of alcoholism volunteered that their "withdrawal" reactions were more severe after stopping aspartame products than after abstaining from alcohol.

- The aggravation of "low blood sugar attacks" by aspartame-containing products can result in a vicious cycle. It is my experience that *most* alcoholics are subject to severe reactive hypoglycemia (Roberts 1971b). This explains in part their craving for sugar and sweets. The alcoholic who consumes aspartame therefore becomes vulnerable to several mechanisms for hypoglycemia—aside from poor nutrition and the impairment by alcohol of enzymes in the liver that transform glycogen (stored starch) into "body sugar."
- The preponderance of females among aspartame reactors (Chapter 8) assumes considerable clinical significance because (1) two percent of women in the United States at any time have a significant problem with alcoholism, (2) the caloric content of alcohol contributes to weight gain, and (3) the use of considerable aspartame for weight control invites serious complications.
- The intake of methyl alcohol, as derived from aspartame, could adversely affect alcoholic patients being maintained on disulfiram (Antabuse). (An estimated 400,000 persons take Antabuse at any given time; a comparable number use the less expensive generic brands of disulfiram.) This drug slows the breakdown of methyl alcohol.
- Some individuals with a previous history of alcoholism appear to have an unusual vulnerability to aspartame products, as the following case reports indicate.

Representative Case Reports

A 47-year-old woman was seen by me in consultation for increasing complaints over an 18-month period. During this time, she consumed large amounts of aspartame in coffee, as diet soft drinks (up to 12 glasses or cups a day), and as puddings. She had had a serious problem with alcoholism 20 years previously, after which she joined Alcoholics Anonymous. She had successfully abstained from alcohol at least five years, and was now happily married.

The patient's symptoms included increasing confusion and memory loss, severe headaches, difficulty in hearing, depression with suicidal thoughts, attacks of "nervousness," intense hunger, a craving for sugar and sweets, muscle cramps, unexplained limb and joint pains, and "dry eyes" (requiring one bottle of artificial tears a week).

The blood pressure was elevated at 130/102. A glucose tolerance test revealed an initial diabetic response, followed by an attack of intense weakness, headache and hunger at the fourth hour, necessitating termination of the study.

On an appropriate anti-hypoglycemia diet and abstinence from aspartame products, her symptoms improved dramatically to the point she was able to resume volunteer relief work within one month. Her blood pressure normalized. Artificial tears no longer were needed. The patient stated that her "withdrawal symptoms" following aspartame abstinence were far worse than when she stopped alcohol.

* * *

A 52-year-old woman with prior alcoholism had abstained from alcohol after 1976. This patient experienced hypoglycemic attacks. Within one to two hours after drinking even a modest amount of aspartame beverages, ear symptoms, drowsiness, slurred speech, marked depression, a change in personality, and intense thirst developed. She wrote

"I reacted to aspartame from my first or second ingestion of the substance. The reaction was so dramatic that it took only four to five instances for me to determine the cause, and halt such intake . . . I felt with aspartame in my system just like I did when drinking alcohol years ago . . . like my brain was drugged. I felt hopeless and helpless . . . It was totally like my drinking days; complete personality change; frightening . . .

"Having suffered from active alcoholism and knowing now the three-part disease I believe it to be (mental, physical and spiritual), as well as the effect on me if I were to take that first drink, I wonder about how many recovering alcoholics have gotten back into problems and/or active drinking after ingesting aspartame and suffering the reactions I did."

CHAPTER TWENTY-EIGHT

The Elderly

(Is There an Aspartame-Alzheimer Connection?)

*Some scholars believe that the downfall of the
Roman Empire just might have been caused by an
overabundance of lead in their systems from their
cooking utensils, etc. I hope we are not seeing the
equivalent of this today in America with all the
unnatural chemical substances we are exposed
to . . . including aspartame.*
— *M A R Y   N A S H   S T O D D A R D*
(President, Aspartame Consumer Safety Network)

MANY HEALTH-CONSCIOUS OLDER persons deliber-
ately consume aspartame products to avoid sugar. Unfortunately,
reactors to aspartame-containing products in this age group face
additional risks.

- The occurrence or aggravation of memory loss related to
  aspartame usage may be misinterpreted as "aging," "small
  strokes," or even Alzheimer's disease (see below) in persons
  over 60.
- Aspartame-associated confusion and drowsiness that contributes to a traffic accident could lead to the loss of an older
  driver's license . . . and severe social and economic hardship.

- Visual problems attributed to cataracts that previously had not caused symptoms could lead to premature surgery (Chapter 14).
- Existing medical problems might be aggravated. Examples include the loss of diabetes control, the aggravation of hypoglycemia, the aggravation of "dry eyes", and the intensification of "dry mouth" with associated difficulty in swallowing and local infection. An important aspect of the related Sjögren syndrome (Chapter 14) is impaired brain function ranging from forgetfulness to dementia (Alexander 1986).

The contributory role of aspartame products in each of these situations has been convincingly shown by (1) dramatic clinical improvement after avoiding these products, and (2) prompt recurrence of the problems on resuming them.

A 66-year-old retired male drank up to eight cans of a diet cola daily for one year. He developed "procrastination, memory loss, bad attitude about myself, itchy skin, headaches, depression, and inability to coordinate my thought." Other complaints consisted of slurred speech, frequent urination (day and night), and suicidal thoughts. These symptoms improved within one week after avoiding aspartame beverages. He stated

"I did notice changes taking place, but I attributed them to the aging process, and even commented that I was perhaps going through 'male menopause.' Then one night, I was watching TV when an expose on aspartame was being aired. The symptoms narrated were just the things I was experiencing. I stopped using the product at once, and some of my problems started to clear up."

A 71-year-old "health conscious" woman remained active and enjoyed fishing. On several occasions after using one packet of an aspartame tabletop sweetener, she developed loss of memory, headaches, intense drowsiness (even while driving), "loss of energy," unsteadiness, unexplained chest pains, heartburn, severe nausea, itching, "open sores," and "dry eyes." She explained

"My judgment was affected a number of times while driving. I would drive across the center line without being aware of it until

I would notice an approaching car. I couldn't understand what was happening to me. I have always been an excellent driver. Once I drove into a ditch and was not conscious of actually doing it. The sleepiness and feverishness increased to the point where I had to lie down every day! I am an avid fisherman, and the thought of not being able to go fishing was very disturbing to me."

After chancing upon an article dealing with aspartame reactions, she eliminated this product. Striking improvement ensued in three days. Her sores healed within three weeks. She described her reaction to a subsequent rechallenge with an aspartame beverage in these terms:

"I discovered a can of sugar-free soda in my refrigerator about three weeks later. Being thirsty and thinking that one can wouldn't affect me, I drank it. About 20 minutes later, I became so sleepy that I had to lie down. There is no doubt in my mind that aspartame is the culprit. If aspartame affects me so drastically, there must be thousands of other people who are affected and don't expect the source to be aspartame. It frightens me to think of the serious accidents I could have been involved in as a result."

AN ASPARTAME-ALZHEIMER CONNECTION?

Many of the clinical sequences described in this book include memory loss and confusion. This raises concern that aspartame consumption—especially when taken in large amounts and over prolonged periods—might precipitate or aggravate Alzheimer's disease and related dementing disorders. The dramatic increase of older persons in our society adds an element of urgency to the clarification of this matter.

On the basis of existing knowledge, Alzheimer's disease should not be considered as either an inevitable age-related loss of neuronal function or a genetic disorder. For example, St. George-Hyslop and colleagues (1987) could not find chromosomal defects in cases of Alzheimer's disease that were sporadic or seemed to affect multiple family members.

There is increasing belief that "benign senescent forgetfulness" and the more serious senile dementia of the Alzheimer's type (SDAT) represent a continuum . . . possibly reflecting a single underlying process. This perspective assumes added significance in the present context since aspartame may accelerate the process.

The *scientific grounds* for such an association include the following observations.

- Procter et al (1986) at the London Institute of Neurology reported a striking loss of aspartate nerve endings in certain important brain cells of patients with Alzheimer's disease.
- Describing research in memory loss and related areas by its scientists, G.D. Searle & Company published full-page ads in leading medical journals (e.g., December 1987 issues of the *Journal of the American Association*) stating that "excessive activity of the excitatory amino acids might cause nerve cell destruction resembling the changes noted in Alzheimer's disease, Huntington's chorea, and other degenerative diseases." (Remember that amino acids of this type are products of aspartame metabolism.)
- Impressive biochemical information pertaining to my observations and inferences is the observation that the amyloid in both the brain plaques and neurofibrillary tangles from patients with Alzheimer's disease contains both aspartic acid (61 per cent) and phenylalanine.

Supporting this conviction, Brayne and Calloway (1988) asserted that the basic questions for future researchers about Alzheimer's disease should be "How much of it has he got?" and "Why?" On the basis of my studies, I suggest that serious consideration be given to the possibility of insidious "early-onset" senile dementia being transformed to Alzheimer's disease by some chemically related environmental neurotoxin, especially in the food supply. Here are just a few related observations.

- Older persons seem to prefer higher concentrations of a mixture of essential amino acids (Murphy 1987).
- Impairment of liver or kidney function in the elderly is conducive to an excessive elevation of both blood and brain phenylalanine levels.
- The ingestion of considerable D-aspartic acid in aspartame products, resulting from excessive heat or prolonged storage,

could have serious consequences. Man and associates (1983) were able to correlate the conversion of the biologically common L (levo) configuration of aspartic acid to its uncommon D (dextro) configuration with aging. They further reported an age-related accumulation of the D-aspartic acid isomer in the myelinated white matter of humans who had no demonstrable brain abnormalities at autopsy. These investigators concluded that this change ". . . may have a bearing on dysfunction of the aging brain or on other diseases of myelin."

CHAPTER TWENTY-NINE

Patients with Phenylketonuria (PKU) and PKU Carriers

The chemicals we ingest may affect more than our own health. They affect the health and vitality of future generations. The danger is that many of these chemicals may not harm us, but will later do silent violence to our children.
— SENATOR ABRAHAM S. RIBICOFF (1971)

PHENYLKETONURIA (PKU) IS AN inherited disease. If undetected and untreated during infancy (and even during pregnancy) children may develop severe mental retardation, convulsions, other neurologic complications, and behavioral disorders.

The cause of PKU is well understood. Affected individuals cannot metabolize phenylalanine normally because phenylalanine hydroxylase, the key enzyme, is severely reduced ... causing excessive accumulation of phenylalanine in the body. As mentioned throughout this book, half of the chemical is phenylalanine and, in my view, many reactions to aspartame products can be attributed to high levels of phenylalanine in the brain and other tissue ... just as in the case of PKU.

Genetic Considerations

PKU occurs in the offspring of two seemingly-normal parents, both of whom carry this single gene defect. Geneticists refer to it as an "autosomal recessive" trait. These *carriers* (or *heterozygotes)* show higher and more prolonged blood concentrations of phenylalanine when given a standard load of this amino acid (Chapter 5).

At least *one out of 50 persons* carries the PKU gene. Accordingly, an estimated four million persons are carriers! The vast majority, however, have no reason to suspect this until the birth of a PKU child.

Patients with PKU and PKU carriers are likely to be at higher risk from consuming aspartame-containing products because of the considerable amounts of phenylalanine. For example, one liter of most aspartame-sweetened soft drinks contains about 275 mg phenylalanine. This explains the mandated warnings on labels—generally in fine print—such as "Phenylpyruvics: Contains Phenylalanine".

Lack of Information

There are serious related problems relative to PKU and aspartame, which reflect a lack of awareness—or interest—of the risks involved.

- The FDA does not require information on labels oriented to the several million female PKU carriers in our population.
- The blood phenylalanine concentrations of female PKU carriers may increase fourfold when they ingest aspartame during pregnancy. (Fetal concentrations are even higher during this critical period of brain development.) Dr. Reuben Matalon (Professor of Pediatrics and Genetics at the University of Illinois Medical School) told a Senate hearing on August 1, 1985 that one in 50 women are "particularly sensitive to high phenylalanine consumption. Their ingestion of aspartame during pregnancy might cause birth defects such as mental retardation."
- Pediatricians and geneticists heretofore have not routinely advised parents and siblings of PKU patients to limit *their* intake of phenylalanine.
- Most aspartame reactors who answered the questionnaire

survey (Chapter 8) knew *nothing* about PKU. Only four individuals were aware of its possible existence in their families.

The possibility of a relationship between the carrier state of PKU and reactions to aspartame-containing products deserves to be explored more fully. In this series, for example, three members of one family with a history of PKU had reactions while consuming large amounts of aspartame products.

- The *husband* developed severe gastrointestinal changes and a presumed "immune deficiency problem" as manifested by alterations in his red blood cells, white blood cells and platelets.
- His *wife* complained of decreased vision in both eyes, severe headaches, extreme dryness of the mouth, and intense thirst.
- Their 28-year-old *granddaughter* had PKU at birth and exhibited learning problems. She experienced multiple grand mal convulsions while consuming aspartame products. Concomitant complaints included headache, mental confusion, memory loss, recurrent depression (with suicidal thoughts), a bleeding peptic ulcer, and anorexia accompanied by a 15-pound weight loss.

Representative Case Report

A 36-year-old woman employed as a "learning-disabled" worker had been diagnosed at birth as having PKU. Her social worker offered these observations relative to the patient's use of aspartame soft drinks.

"Aspartame is like a poison to her . . . but it is very difficult to avoid aspartame . . . It is everywhere in diet colas, diet root beer, sugar-free drinks and gelatins. She can't sleep so that she can go to work. It takes 5 days before her system is back to normal . . . My reward comes when she comes in from work and gives me that bone-crusher hug around the shoulders and says, Boy, do I feel good."

Her symptoms following aspartame intake consisted of severe headache, dizziness, unsteadiness, mental confusion, memory loss, marked hyperactivity, insomnia, slurred speech, attacks of shortness of breath, palpitations, abdominal pain, severe nausea, diarrhea with bloody stools, and abdominal bloat. The social worker added

"Her reaction is so severe that I have stayed up with her 2 nights for fear she might pass out or her heart give out . . .

"All we could do for months was to restrict foods that made her sick as each one showed itself. (She was living on junk food and lived in the manner of a street bag lady before this) . . . She was tested at a medical center for allergies. Three trays of 18 were tried on her . . . Dr. ——— says her problems are caused by the PKU factor."

CHAPTER THIRTY

Does Aspartame Pose an Imminent Public Health Hazard?

Human history becomes more and more a race between education and catastrophe.

— H. G. WELLS

THE DRAMATIC INCREASE IN the consumption of products containing aspartame demands careful reevaluation of the potential adverse effects of this additive on the population. A report on the Food Additives Amendment for the House of Representatives stated: "Safety requires proof of a reasonable certainty that no harm will result from the proposed use of the additive" (H.R. Report No. 2284, 85th Congress, 2nd Session, 1958). Is there "reasonable certainty" that no harm results from aspartame? Not in my view.

The apparent flaws in animal studies prior to marketing, the absence of extensive testing of aspartame-containing products on humans prior to licensure, lax standards for the approval and surveillance of food additives, and the escalating number of serious aspartame-associated complaints *volunteered* by consumers warrant this opinion.

Yet, governmental agencies charged with investigating these matters continue to express minimal concern. For example, there is no "early warning system" for reporting severe reactions from food additives "generally regarded as safe" (GRAS) comparable to that required for over-the-counter drugs. Post-marketing surveillance is even worse in other countries.

The potential toxicity of aspartame was inferred as early as 1970 when Cloninger and Baldwin suggested that ". . . by maintaining low concentrations of sweetening agents, problems of toxicity are less likely." Shortly after the release of aspartame-containing products, Uribe (1982) criticized the FDA for not having carefully investigated its toxicity in patients with liver disease.

The Limitations of Animal Studies

But how can this potential danger be determined? *Remember, there is no reliable animal model for evaluating aspartame toxicity in humans.*

- The rat liver metabolizes phenylalanine five times more efficiently than the human liver. Therefore, the results cannot be compared meaningfully.
- Even most monkey species are able to detoxify methyl alcohol, unlike the severely limited ability of humans to do so.
- Another shortcoming of using rats and other experimental animals is their apparent failure to perceive aspartame as "sweet." Studies with such animals therefore do not replicate the reflex (cephalic) phase of insulin release triggered in man by the taste of sweet.

Also, many individuals who react to aspartame products have other existing conditions which increase the risk of toxicity—circumstances quite different than in animals tested under strict protocols and optimum laboratory conditions.

In high-risk group are pregnant and nursing mothers, infants and young children, older persons, and individuals having iron-deficiency anemia, liver dysfunction, kidney impairment, migraine, hypoglycemia, diabetes mellitus, alcoholism, and phenylketonuria (PKU). Furthermore, patients taking drugs that might interact with phenylalanine—for example, L-dopa, methyldopa, and monoamine oxidase (MAO) inhibitors—are included in the general population, but not in "controlled" human and laboratory studies.

Bailar and colleagues (1989) questioned the confidence of risk management from exposure of consumers to substances in foods and drugs that may cause cancer on the basis of experimental studies involving "genetically uniform laboratory animals, maintained under uniform conditions and protected from possibly damaging exposure to other agents."

Appellate courts have ruled on the shortcomings of animal studies relative to the effects of drugs and other chemicals in humans. Some examples:

- In *Lynch v. Merrell Dow National Laboratories* (1987), it was ruled that animal studies were found lacking in probative value relative to admissibility in proving human causation because (1) they were performed in different biological species, and (2) doses far greater than the human therapeutic dose were administered.
- It was successfully argued in the extensive Bendectin litigation that animal studies, both *in vivo* and *in vitro*, are less reliable than studies on humans *(Richardson v. Richardson— Merrell, Inc.* 1988).

Recollections of Thalidomide

I hope not, but *ignoring the long-term problems seemingly associated with use of aspartame-containing products might portend another "thalidomide revolution".* It is imperative that researchers who are not influenced by producer–corporations gather detailed data concerning the incidence rates of epilepsy, visual problems, allergies, Parkinsonism, brain tumors and Alzheimer's disease among aspartame consumers. The same applies to other disorders—including the frequency of birth defects, seizures, impaired intelligence and behavioral abnormalities among the children of women who ingested aspartame products at conception and during pregnancy.

The possible emergence of "new and improved" formulations of aspartame products may provide an early clue that these projections are on target. Previous examples of restructuring troubled products include (1) "new filter" cigarettes which were advertised during the early 1960s as "Just What the Doctor Ordered," "Not a Cough in a Carload," and "Play Safe—Smoke Chesterfields," and (2) the introduction of anti-static clothes softeners without perfumes, inks and other additives following reports by the writer (Roberts 1986) and others of severe respiratory and skin irritation.

MAGNITUDE OF THE ASPARTAME PROBLEM

The current situation ought to be regarded as one of *caveat emptor* ("let the buyer beware") as long as the FDA continues to deny the existence of serious problems associated with aspartame use. This is illustrated by the testimony of its Commissioner to a Senate hearing held on November 3, 1987. He reassured the public concerning the safety of aspartame products, notwithstanding his statement that 3,679 aspartame-related complaints had been received from consumers as of October 23, 1987 . . . an unparalleled phenomenon.

It was mentioned earlier that if only *one-tenth of one percent* (0.1%) of the 100 million persons in the United States currently consuming aspartame products developed adverse reactions, the actual number would be *100,000* persons! In my opinion, a far higher figure is more realistic.

Some personal experiences vouch for the widespread existence of aspartame reactors in the *general* population.

- The day after my first interview on a West Palm Beach television station, my office received calls for the survey questionnaire from *28* persons who thought they were probable aspartame reactors.
- The office manager of a large law firm called for 90 (!) forms that same day. I questioned her about this unusual request. The reason: she had observed profound changes in many employees who were consuming large amounts of diet colas—especially depression, erratic behavior and poor work performance. Being in charge of the machines that dispensed these drinks, the manager had an objective basis for making such a correlation.
- After an interview on Radio Station WWDB in Philadelphia—a session that lasted less than one hour—my office (which is in Florida) received *more than 100* letters, cards and phone calls about aspartame.

Far more important is that at least *10,000* persons have registered aspartame-related complaints with governmental agencies, companies and professionals other than myself. Admittedly, some of these consumers may have been in the files of two or more organizations.

- The Food and Drug Administration (FDA) and the Centers for Disease Control (CDC)—over 5,300 complaints.
- G. D. Searle & Company and The NutraSweet Company— over 3,000 complaints (most relayed to the FDA).
- The Community Nutrition Institute—over 2,500 complaints.
- Dr. W. C. Monte (Food Sciences Department, Arizona State University)—over 1,000 complaints.
- Dr. R. J. Wurtman (Massachusetts Institute of Technology)— more than 100 persons with seizures attributed to aspartame, and "well over 1,000" letters and "related communications" (*Congressional Record-Senate* May 7, 1985).
- Aspartame Victims and Their Friends—more than 900 complaints.

Another projection of the scope of the problem is provided by Dr. Anthony Kulczycki, Jr. (1987), who estimated that a minimum of *6,000* persons in the United States are allergic to aspartame products.

THE IMMINENT PUBLIC HEALTH HAZARD ISSUE

The issue of an "imminent public health hazard" has been previously raised relative to drugs and environmental poisons . . . generally on the basis of a few cases. For instance, phenformin (DBI®), a drug that had been used successfully in treating many persons with diabetes mellitus, was declared an imminent public health hazard— and removed from the market—because a rare complication (lactic acidosis) developed in several dozen patients (among many tens of thousands) . . . chiefly because they should have been receiving insulin or another drug.

Aspartame-containing products have managed to evade this onus for various reasons.

- Since it is classified as a *food additive*, aspartame does not come under as intense scrutiny by the FDA as do drugs (Callaway 1986).
- The FDA adamantly asserts that aspartame products are safe for nearly all consumers who do not have PKU (Chapter 1).
- Aspartame products are backed by powerful economic and political interests. This could be inferred from the recorded

contributions received by key senators following important votes favorable to this industry (Gordon 1988).

- Mintz and Cohen (1971) detailed the ability of giant corporations to influence environmental decisions by placing "a discreet phone call" to bureaucratic officials that could "abort a necessary regulatory action" (p. 12).
- Manufacturers and producers are not required to offer evidence for safety and efficacy if they avoid specific therapeutic claims (such as weight loss) on the package label.
- Physicians are not mandated by law to report adverse reactions to food additives.

The Issue of Ultimate Proof of Causation

Citizens of the United States take proper pride in a judicial system that assumes an accused individual to be innocent until proved guilty by a jury of peers. However, when corporate attorneys demand "ultimate and absolute proof" of causation for the alleged toxicity of a substance (such as aspartame) being consumed by nearly half the population, and in the face of thousands of *volunteered* adverse reactions to it, appellate courts must not allow the scale of justice to be tipped away from millions of innocent consumers. Similarly, without some legislative or executive restraint, many more are likely to risk serious health problems as "battles of the experts" drag on for months or years . . . given the almost unlimited funds for legal defense available to corporate vested interests.

An expert medical witness must recognize the limits of his testimony "with reasonable certainty." The insistence by either opposing counsel or a court for definitive and absolute proof of causation, especially in toxic tort litigation, may be unreasonable. This dilemma is especially acute when an opinion or theory is based on unorthodox, "well-founded methodology."

The expert medical witness necessarily deals within the realm of "scientific proof." However, both he and the trier of fact must recognize the lack of ultimate and absolute answers, especially in clinical and epidemiological studies.

- Bradford Hill (1965) stressed the incompleteness of all scientific work, whether observational or experimental, and its susceptibility to modification by newer knowledge.

- Dr. Jerome P. Kassirer (1989) aptly editorialized on the unobtainability of "absolute certainty in diagnosis." In spite of the number of observations, tests, and other information gathered, he asserted that the task of the clinician ". . . is not to attain certainty, rather to reduce the level diagnostic uncertainty enough to make optimal therapeutic decisions."

The issue of "statistical significance" also is germane. It was considered by a United States District Court in *Nehmer v. United States Veterans Administration* (1989). This case represented another round in the conflict between Vietnam veterans and the United States Government over alleged debilitating diseases resulting from exposure to Agent Orange. The Court stated:

The VA wrongfully required that the scientific evidence demonstrated a "cause and effect" relationship between Agent Orange exposure and claimed diseases, instead of using the less demanding standard that there be a "statistical association" between Agent Orange exposure and claimed diseases . . . A statistical association "means that the observed coincidence in variations between exposure to the toxic substances and the adverse effect is unlikely to be a chance occurrence or happenstance."

The Imminent Hazard Regulation

There is a pertinent FDA regulation (21 CFR 2.5) concerning an imminent public health hazard.

"(a) Within the meaning of the Federal Food, Drug and Cosmetic Act, an imminent hazard to the public health is considered to exist when the evidence is sufficient to show that a product or practice, posing a significant threat of danger to health, creates a public health situation (1) that should be corrected immediately to prevent injury and (2) that should not be permitted to continue while a hearing or other formal proceeding is being held. The 'imminent hazard' may be declared at any point in the chain of events which may ultimately result in harm to the public health. The occurrence of the final anticipated injury is not essential to establish that an 'imminent hazard' of such occurrence exists.

"(b) In exercising his judgment on whether an 'imminent hazard' exists, the Commissioner will consider the number of injuries anticipated and the nature, severity, and duration of the anticipated injury."

Various licensed products and drugs have been removed by the FDA and the Environmental Protection Agency (EPA) in the face of evidence they constituted an imminent public health hazard.

- The FDA banned the use of sulfite preservatives on fresh fruit and vegetables during August 1986.
- Only several dozen complaints were sufficient to remove certain liquid protein formulas advocated for weight loss, and formaldehyde from home insulation.

The Concept of Food Additives Safety

The Congress approved a food additive amendment in 1958. It was intended to protect the public against unsafe food additives to the same degree as unsafe drugs.

This objective was stated in *Congressional and Administrative News. Legislative History of Food Additives Amendment of 1958* (p. 5302): "Safety requires proof of reasonable certainty that no harm will result from the proposed use of an additive." It further amplified the "Concept of Safety" for food additives in these terms:

"The concept of safety used in this legislation involves the question of whether a substance is hazardous to the health of man or animal. Safety requires proof of a reasonable certainty that no harm will result from the proposed use of an additive. It does not and cannot require proof beyond any possible doubt that no harm will result under any conceivable circumstance.

"This was emphasized particularly by the scientific panel which testified before the subcommittee. The scientists pointed out that it is impossible in the present state of scientific knowledge to establish with complete certainty the absolute harmlessness of any chemical substance.

"In determining the 'safety' of an additive, scientists must take into consideration the cumulative effect of such additive in the diet of man or animals over their respective life spans together with any chemically or pharmacologically related substances in such diet. Thus, the safety of a given additive involves informed judgments based on educated estimates by scientists and experts of the anticipated ingestion of an additive by man and animals under likely patterns of use.

"Reasonable certainty determined in this fashion that an additive will be safe, will protect the public health from harm and will permit sound progress in food technology.

"The legislation adopts this concept of safety by requiring the Secretary to consider in addition to information with regard to the specific additive in question, among others, the following relevant factors: (1) the probable consumption of the additive and of any substance formed in or on food because of the use of such additive; (2) the cumulative effect of such additive in the diet of man or animals, taking into account any chemically or pharmacologically related substances in such diet; and (3) safety factors which qualified experts consider appropriate for the use of animal experimentation data.

"In determining the safety of an additive, the Secretary would have to consider not only the food to which the additive, is directly added, but also other foods derived from such foods. For example, in evaluating the safety of an additive for poultry feed, the Secretary would have to consider any residues that might appear in eggs produced by the poultry. Similarly, in determining the safety of additive-treated cattle feed, account would have to be taken of residues of the additive in the milk or edible flesh of the animal.

"Since the scientific investigation and the other relevant data to be taken into consideration by the Secretary include information with respect to possible cancer causing characteristics of a proposed additive, the public will be protected from possible harm on this count."

Bureaucratic differences of opinion, however, have surfaced. Some of the criteria used by the Bureau of Foods for additive safety are not acceptable to the Bureau of Drugs.

ASPARTAME AS A "DRUG"

Aspartame, as noted, has significant biological effects. Accord-

ingly, there are legitimate grounds for reclassifying it as a "drug."
If this happened, manufacturers, packagers and distributors would
be required to report "serious and unexpected" adverse reactions
and "any significant increase in the frequency" of expected or serious
reactions within 15 working days.

Over-the-Counter Phenylalanine

The alleged value of phenylalanine (constituting half of the aspartame molecule) and other pure amino acids (such as tryptophan,
tyrosine, lysine and ornithine) in restoring energy, combating depression, improving memory, aiding alcoholism, minimizing stress,
and promoting weight loss is constantly extolled by the "wellness"
press. Various evidences are offered as scientific rationale for these
claims. In the case of L-phenylalanine, they include (1) its conversion
to tyrosine—and then to dopamine and norepinephrine— for an antidepressant effect, (2) its decarboxylation to phenylethylamine (a
compound having chemical similarities to, and behavioral effects resembling, amphetamine) for combating fatigue, and (3) the inhibition
of metenkephalon degradation for the relief of pain. Not surpris-
ingly, these amino acids are popular items in "health stores."

The use of these pure amino acids is potentially dangerous in view
of their pharmacologic potential for altering brain and neurotransmitter function (Chapter 5). This is supported by experiences of
patients and correspondents. Some aspartame reactors reported that
their reactions were more severe to phenylalanine itself than to aspartame-containing products.

> A California woman wrote me concerning her side effects from
> phenylalanine and tyrosine. Although she had enjoyed "excellent
> health my whole life," a health food clerk recommended two
> grams a day of phenylalanine and tyrosine. She experienced pro-
> gressive "memory loss, dizziness, inability to concentrate, shoot-
> ing pains in the left arm and leg, marked sensitivity of the eyes
> to light, and extreme cold in the left limbs."

> In addition to consuming considerable amounts of aspartame-
> containing beverages, presweetened tea and diet cranberry juice,
> one young woman with severe gastrointestinal complaints, head-
> aches, memory problems and anxiety also was taking phenylala-

nine, arginine and ornithine to lose weight. (She weighed only 133 pounds.)

Other investigators share my concern about these amino acids. Furthermore, the general failure of L-phenylalanine or D-phenylalanine to ameliorate depression and pain led the Canadian government to prohibit their sale (Young 1987).

CHAPTER THIRTY-ONE

Public Health Issues

*Long-delayed effects are frequently difficult to
relate to their specific causes. Thus, it is
conceivable that injuries due to unrecognized
causes have been produced by common materials
that have long been considered safe as food or for
use in foods. In this regard we are concerned with
cancer, genetic damage and birth defects,
premature old age, cardiovascular, endocrine and
mental disorders, and other human ills of
unexplained etiology.★*
— D R. J ULIUS M. C O O N (1970)

REACTIONS TO ASPARTAME-CONTAINING products
may pose potential public health threats. This chapter will address
some of these possibilities—including aspartame-related accidents,
suicide, criminality, cancer, and excessive diagnostic radiation.

ASPARTAME-RELATED ACCIDENTS

Aspartame reactions that affect vehicle drivers, pilots, air traffic
controllers, nuclear plant operators, and personnel engaged in other
strategic occupations could result in or contribute to serious accidents. Such individuals often consume large amounts of sweetened
coffee, tea and cola beverages in an attempt to remain alert. In this

★ © 1970 *Modern Medicine*. Reproduced with permission.

current era of health-consciousness, many are also likely to use aspartame products instead of sugar.

Aspartame-related mental confusion, loss of memory, fatigue, visual problems, depression, sleepiness, convulsions, changes in behavior, and heart rhythm disturbances are highly relevant. Case histories reported earlier include the grounding of commercial or military pilots as a result of seizures after consuming aspartame products.

In my book on traffic accidents (Roberts 1971b), I have detailed the deterioration of driving and flying skills from subtle physiological alterations. The current estimate of 2,000–3,000 near-collisions in the air annually (*The Wall Street Journal* July 21, 1987, p.33) is pertinent.

> The captain of commercial 747 planes related his reactions to aspartame products. They consisted of depression, altered behavior culminating in a recent divorce, and deterioration in flying skill perceptible to him. The adverse effects on performance became particularly apparent when there were lengthy delays before departure or landing. He felt markedly improved after avoiding such products. This senior pilot then expressed concern about similar reactions among his colleagues who were ingesting large amounts of aspartame products.

SUICIDE AND CRIMINALITY

Changes in mood, personality and behavior have been extreme among some reactors to aspartame products (Chapter 13). The ramifications of severe irritability, depression with suicidal thoughts, and aberrant behavior conducive to criminality merit particular attention among children and teenagers consuming considerable amounts of these products.

The precipitous rise of bizarre homicides and highway violence is relevant. "Road wars" and "copycat shootings" have become commonplace on Southern California's overcrowded freeways. Law enforcement officials remain mystified over the inappropriate fist fights, shootings, and other forms of mayhem committed by otherwise responsible persons after some minor traffic encounter (e.g., failure to yield, or the flashing of high beams by an oncoming car) (*The Wall Street Journal* August 3, 1987, p.1).

My wife is one of five elected Commissioners of Palm Beach County. She made an interesting observation while surveying Florida's county and state prisons during April 1989: the presence of large supplies of an aspartame tabletop sweetener. This evoked the question, "If aspartame *can* cause mental and behavioral abnormalities in otherwise presumably normal individuals, what might it do to convicted felons?"

CONCERN OVER CANCER

There has been virtually no outcry by health professionals, the FDA or the public over the approval of aspartame-containing products for human consumption when pre-licensing animal studies showed brain and uterine tumors in rats (Cornell 1984). This information appears in the *Congressional Record—Senate* hearings of May 7, 1985 (pp. S5489–5516) and August 1, 1985 (pp. S10820–10847).

Objections By Other Investigators

Dr. M. Adrian Gross, then a senior FDA pathologist, told the Senate hearing held on August 1, 1985

"In view of all these indications that the cancer-causing potential of aspartame is a matter that has been established way beyond any reasonable doubt, one can ask: 'What is the reason for the apparent refusal by the FDA to invoke for this food additive the Delaney Amendment to the Food, Drug, and Cosmetic Act?' "

Dr. Douglas L. Park, Staff Science Advisor for the Office of Health Affairs of the Department of Health & Human Services, concluded his analysis of the hearing by a Public Board of Inquiry (PBOI) on aspartame safety in 1981 relative to brain gliomas in treated rats in these terms:

"I believe that aspartame has not been shown to be safe for the proposed food additive uses. Along with the Board of Inquiry, I must recommend, therefore, that aspartame not be approved until additional studies are carried out using proper experimental designs."

Dr. John Olney, Professor of Psychiatry and Neuropathology at Washington University School of Medicine, wrote the following statement to Senator Howard Metzenbaum, dated December 8, 1987, concerning aspartame-related brain tumors.

"This is an exceedingly complex topic which, unfortunately, has a history riddled with appearances of fraudulent practices by the manufacturer of NutraSweet and ineptitude and/or malfeasance on the part of the FDA officials. In the mid 1970's, when I reviewed the NutraSweet record in preparation for the hearing I had been promised, I came upon a peculiar study which the manufacturer had submitted to the FDA and which FDA had unquestioningly accepted as evidence for the safety of Nutra-Sweet. The study showed that in 320 NutraSweet-fed rats there were 12 brain tumors whereas in a group of concurrent control rats which were not exposed to NutraSweet, there were no brain tumors. Being a neuropathologist, I know that spontaneous brain tumors in laboratory rats are extremely rare. The archival literature documents an incidence not exceeding 0.6%. Since the above incidence in NutraSweet-fed rats is 3.75%, this suggests that NutraSweet may cause brain tumors and certainly suggests the need for additional in-depth research to rule out that possibility . . .

"I seriously doubt whether this method of data analysis would stand the scrutiny of competent disinterested statisticians. Even more seriously I wonder why FDA allows microscopic slides to disappear (while supposedly impounded) and why they do not question the *de novo* emergence of a brain tumor among the controls when the slides reappear.

"The PBOI panel member who was primarily responsible for reviewing the brain tumor issue was Peter Lampert, M.D., Neuropathologist and chairman of the pathology department at Univ. of Calif. San Diego. Dr. Lampert personally examined the microscopic slides pertaining to the brain tumor studies and told me a year or so after the PBOI report was completed that he had been surprised at the large size of the brain tumors in the NutraSweet-fed rats."

Senator Metzenbaum offered this commentary at the May 7, 1985 Senate hearing:

"I do not claim children will develop brain tumors. I do not know that. I do know that the FDA was worried about it. I do know that three of the six FDA scientists advising the FDA Commissioner on final approval were sufficiently worried about it that they were not willing to approve the product. The FDA's own scientists were split on the issue."

Other observations in both animals studies and aspartame reactors should evoke concern about *the potential tumor-causing role of aspartame-containing products.*

- The excision of tissue masses from live rats during pre-licensing studies according to the so-called Bressler report.
- The failure to report such presumed tumors to the FDA.
- The finding of uterine polyps and ovarian tumors—also not reported to the FDA.
- The occurrence of blood changes resembling leukemia, and of lymph node enlargement, among persons consuming aspartame. (These data in my series have not yet been published.)

The Issue of Gender

As far as I have been able to determine, the agencies responsible for licensing aspartame-containing products overlooked at least one important detail in assessing the animal studies: *the carcinogenic risk of aspartame and its diketopiperazine derivative (Chapter 5) was analyzed ONLY in FEMALE mice!* Professor George T. Bryan (1984) stated that *female* Swiss albino mice were utilized for "all" such studies.

This observation may be important because a gender vulnerability has been noted in saccharin studies (see below). Miller and Howe (1977) aptly noted: " . . . for bladder cancer the distinction between males and females seems to us to be fundamental."

The Delaney Amendment

It is proper to ask the question: "Why wasn't the Delaney Amendment invoked for aspartame use?" This clause bans cancer-promoting chemicals from foods. It was incorporated by Congress in the 1958 Food Additives Amendment.

There have been recent arguments in favor of a *de minimis* interpretation of the Delaney clause by the Food and Drug Administration (FDA). The provision in the Food, Drug and Cosmetic act prohibits that agency from approving the use of cancer-causing substances in food, drugs and cosmetics. The Congress has been sufficiently clear in the matter that the FDA cannot rewrite the law unilaterally (Schultz 1989). This precaution is particularly pertinent when it is likely the risk estimate has been grossly underestimated, especially in animal studies that focus on only one variable (the chemical itself), rather than the combination of potential carcinogenic risks encountered by the consuming public. William B. Schultz (1989) pointed out that a risk of 1 in 9,000,000 could translate to a risk of 1 in 90,000 in the case of exposure to 100 chemicals, and to 1 to 9,000 in the case of 1,000 chemicals.

When this is the basis for concern about the toxicity and carcinogenicity of a food additive, for example, courts must consider arguments about the lessened availability of a variety of foods in proper perspective. They also must recognize that "risk management" still remains more an art than a science (Bailer 1989).

The Saccharin Controversy

For several decades, I have monitored the highly controversial allegation that saccharin induces urinary bladder tumors in rats. It is pertinent to discuss this subject because the public perceives saccharin as unsafe due to the mandated labeling of foods and drinks containing it . . . with the inference that saccharin causes cancer. As a result, consumers prefer "safer" products—such as those sweetened with aspartame.

The facts of this "mouse-to-man" controversy are summarized.

- The alleged carcinogenicity of saccharin basically derives from finding of urinary bladder tumors in *a few* rats that had been given *large* amounts of the sweetener.
- The initial studies by the Wisconsin Alumni Research Foundation (1973), suggesting a statistically significant increase of bladder tumors, indicated that such tumors occurred *only* in *male* rats (see above).
- The use of rats for these pellet-implantation studies has come under severe criticism. Dr. Freddy Homburger (1977), an

authority on the experimental use of Sprague-Dawley rats, clearly asserted: "Rats are not appropriate animals for the study of bladder cancer."

- After a review of several large human epidemiologic studies, the President of the American Cancer Society stated: "There is no evidence that saccharin causes cancer in humans" (*Science* 196:276, 1977).

- Following years of intensive research, Dr. Frank Young, FDA Commissioner, concurred that no link had been proved between saccharin consumption and the development of cancer in humans (*Congressional Record—Senate* May 7, 1985, p. S5512).

- A recent position statement on saccharin by the American Diabetes Association (1987) concluded: ". . . epidemiological studies show no evidence of a carcinogenic effect (of saccharin) in humans . . . the small risk, if any, from ingestion at moderate levels is considered to be extremely small."

The Cyclamate Controversy

In a similar vein, controversial research led to the withdrawal of cyclamate from the United States market in 1970. Nevertheless, this artificial sweetener is permitted in low-calorie foods by 40 countries, including Canada.

Cyclamate, discovered in 1937, is approximately 30 times sweeter than table sugar (sucrose). It has the added advantages of being heat-stable and devoid of saccharin's aftertaste.

Specifically, a study with rats given a saccharin/cyclamate mixture suggested that it might cause cancer in this rodent (Price 1970). Yet, in a summary statement on various sweeteners, dated August 1986, the Institute of Food Technologists Expert Panel on Food Safety & Nutrition indicated that subsequent research *failed* to prove conclusively any tumor-producing properties of cyclamate.

The rejection by the FDA in 1980 of a petition for approval of cyclamate was based on the general safety clause of the Food, Drug and Cosmetic Act . . . not the Delaney Clause. In the absence of such data and evidence for any mutagenic effect of cyclamate, additional petitions were introduced. Even the FDA's Cancer Assessment Committee of the Center for Food Safety and Applied Nutrition asserted in April 1984 that no further research into the question of cyclamate carcinogenicity was necessary!

NEEDLESS EXPOSURE TO RADIATION AMONG ASPARTAME REACTORS

The extensive and repeated exposure of aspartame reactors to numerous diagnostic x-rays is reflected in many of the case histories presented earlier. Scans and other studies of the brain are now virtually routine for patients presenting with neurological complaints. The same applies to radiographic studies of the stomach, small intestine and large bowel when peptic ulcer, Crohn's disease and ulcerative colitis are suspected in persons with aspartame-related gastrointestinal complaints.

The frequency with which young persons underwent such needless (and costly) exposure to radiation is disturbing, especially when the possibility of an aspartame reaction was seldom considered.

SYNTHETIC ADDITIVES AND FOOD SUBSTITUTES: POTENTIAL MEDICAL AND NUTRITIONAL PROBLEMS

Physicians, dietitians and regulatory bodies must be reflexively triggered to the public health implications of unnatural substances (referred to as xenobiotics), especially when synthetic. The "technological tweak" given Nature by "designer" foods and additives as a result of achievements by agricultural scientists should be of *extraordinary* concern to regulatory and licensing agencies.

The aspartame experience reported in this book underscores the potential hazards associated with *the uncommon D-stereoisomers of amino acids* resulting from alteration of the common L-stereoisomers during food processing, synthesis and preparation (Chapters 5). They include impaired digestibility of such amino acids, their lessened bioavailability, and even interference with the metabolism of biologically active L-amino acids.

The increased sophistication in modifying or substituting foods with *products that have few or no calories* also poses a major concern. The number of synthetic taste-alike products for sugar, fat and protein—or combinations thereof (such as the proposed addition of aspartame to a fat substitute for a simulated ice cream)—is likely to mount rapidly. Furthermore, biotechnologies for synthesizing proteins and incorporating amino acids into genetic material, as with recombinant DNA, now exist.

I do not consider the present standards for evaluating the biophysiologic, neuropsychiatric and behavioral consequences of such products in *humans* to be definitive. A current example concerns the so-called microparticulation of egg and milk protein for making one type of "fake fat" (Roberts 1989a). This state of affairs is superimposed upon the lax regulation of non-food, non-drug additives and substitutes derived from "generally regarded as safe" (GRAS) substances.

Others share these evolving worries (Johnson 1988).

- Individuals concerned about excess weight, especially young women, risk caloric deprivation . . . and the potentially serious consequences thereof (Roberts 1985).
- The population is likely to pay less attention to ensuring a healthy and adequate diet by focusing excessively upon tasty food substitutes.
- Patients having diabetes mellitus, increased cholesterol concentrations, heart disease, and other disorders understandably will indulge themselves with "sugar-free", "cholesterol-free" and "fat-free" substitutes—especially when recommended without reservation by their physicians and dietitians.
- Greater consumption of sugar already has accompanied the increasing use of aspartame and other artificial sweeteners (Chapter 3).
- Laxative products containing aspartame and other substitutes for sugar and fat might result in the malabsorption of food nutrients and fat-soluble vitamins and minerals (Chapter 19).

It would seem axiomatic that these considerations should be addressed *before* the wholesale release of synthetic food substitutes, particularly if large pre-marketing studies on *humans* had not been performed (as in the case of aspartame.) I have repeatedly emphasized (Chapter 30) that animal experiments are *not* reliable for several reasons, including differences in metabolism and life span. (The life cycle of rats or mice is only about four years.)

Judicial rulings concerning the licensing and availability of products to which a large segment of the population might be exposed has been analyzed by the courts. In *Brown v. Superior Court* (1988), the Supreme Court of California restated the so-called Kearl test, wherein a judge should decide after hearing the evidence

"(1) whether, when distributed, the product was intended to confer the exceptional important benefit that made its availability highly desirable; (2) whether the then-existing risk posed by the product was both 'substantial' and 'unavoidable'; (3) whether the interest in availability (again measured as of the time of distribution) outweighs the interest in prompting enhanced accountability through strict liability design defect review." (p.481)

CHAPTER THIRTY-TWO

The Myth of "The Most Thoroughly Tested Additive in History"

*Finally, we should learn a lesson from the
NutraSweet experience. If a food additive has
potential neurological or behavioral effects, it
should undergo human clinical testing, similar to
the process a drug must undergo before it is put on
the market . . . the food and beverage industry,
and their various institutes, exert tremendous
influence over scientific research and investigation.
I want to make sure such work is genuinely
independent.*

— *S E N A T O R   H O W A R D
M E T Z E N B A U M (1988)*
(Hearing on "NutraSweet"—
Health and Safety Concerns")

T HE APPARENT MAGNITUDE OF adverse reactions to
aspartame products, and related public health problems, have been
discussed in prior sections. This chapter summarizes certain shortcomings pertaining to research, labeling, "disinformation," and the
acceptable daily intake (ADI) of such products. It also will review
professional, legislative and bureaucratic obstacles encountered by
physicians, patients and consumer advocates.

These problems are not new . . . only magnified. Sinclair Lewis criticized the commercialism and bureaucratic roadblocks encountered by motivated physicians and researchers in his 1925 novel *Arrowsmith.*

DEFICIENCIES IN LABELING

Food labeling regulations originated with the Federal Food and Drug Act of 1906. It prohibited false or misleading statements on food or drug labels. Dr. Allen L. Forbes (1986), Director of the Center for Food Safety and Applied Nutrition of the FDA, asserted: "Extreme care is needed to maintain the accuracy of such information and the integrity of the food label generally."

I am convinced that *products containing aspartame should contain specific information about the quantity of this additive and a recommended expiration date.* "Buzz words" on labels—such as "sugar-free" and "light"—tend to confuse consumers.

For added perspective, I will call your attention to the situation in Israel. Since 1984, this country has mandated that labels state in both English and Hebrew: "Do not consume more than three grams aspartame daily for adults." The expiration date is generally two months after production. Another precaution reads: "Store in a cool place not over 22° Centigrade." Certainly, it must be assumed that a contemporary nation confronted by many governmental problems and priorities would not take such steps involving widely used consumer products having considerable economic value without the input of much scientific expertise and consultation. (By coincidence, I had lunch with a high-level Israeli public health official at the first international conference on Dietary Phenylalanine and Brain Function held in Washington during May 1987, which his government considered sufficiently important for him to attend.)

Without such specific information on labels, consumers have no way of knowing that (1) certain "diet" colas and sodas (especially orange drinks) contain considerably more aspartame than others, or (2) aspartame products may have undergone drastic chemical alterations during a long shelf life in stores or as the result of a hot environment.

The quantity of aspartame is of particular importance to parents who are concerned about the amounts of this additive their children consume. In this vein, Senator Howard Metzenbaum stated, "We cannot use America's children as guinea pigs to determine the safe level of aspartame consumption" (*The Philadelphia Inquirer* March 4, 1986, p. A–4).

Some patients and correspondents clearly suffered from the erroneous labeling of aspartame products.

> A 66-year-old woman had been in good diabetic control with an appropriate diet and oral medication. After drinking one brand of a diet cola, she experienced "such severe headaches I couldn't function and was disoriented". These reactions promptly occurred on each of *three* retesting trials. She then drank a soft drink labeled as being sweetened with saccharin. Her intense headache recurred after ingesting only one-fourth of the can. ("I felt my brain was swelling (and) to be too big for my cranium.") She stated:
>
> "I called the bottling company and asked if they were using aspartame. They said *yes!* I told them the can said saccharin. They told me they were just using up the old cans, and then they would change the contents as listed. I didn't get any satisfaction from them, so I called the Food and Drug Department in Oahu. I told them the story and they just said, 'Yes, that's what they do.' They were very uncooperative. I felt I was hitting my head against a brick wall. They didn't care a thing about me.
>
> "The outcome is that I eat nothing with aspartame listed as an ingredient. I find very few drinks without it. It is difficult for me, as I am not a coffee drinker."

The issue of proper and adequate labeling is clearly pertinent to the post-sale duty to inform. Several examples not involving aspartame are cited.

- The Supreme Court of New Jersey emphasized in *D'Arienzo v. Clairol, Inc.* (1973) that labels must not be ambiguous for "certain individuals" who may be allergic or hypersensitive to a product even though it seems "harmless to the multitude."
- Congress decided during 1988 that a specific health warning be placed in a "conspicuous and prominent space" on every container of liquor, beer and wine. The wording of such an

advisory under the law must state: "GOVERNMENT WARNING: (1) According to the Surgeon General, women should not drink alcoholic beverages during pregnancy because of the risk of birth defects. (2) Consumption of alcoholic beverages impairs your ability to drive a car or operate machinery and may cause problems."

- In *MacDonald v. Ortho Pharmaceutical Corporation* (1985), the Supreme Judicial Court of Massachusetts emphasized the peculiarly characteristic nature of some products—in this instance oral contraceptives—that warrants the imposition of a common law duty on the manufacturer relative to a direct warning for users . . . even transcending the role of the prescribing physician as a "learned intermediary."

"We conclude that the manufacturer of oral contraceptives is not justified in relying on warnings to the medical profession to satisfy its common law duty to warn, and that the manufacturer's obligation encompass a duty to warn the ultimate user. Thus, the manufacturer's duty is to provide to the consumer written warnings concerning reasonable notice to the nature, gravity, and likelihood of known or knowable side effects, and advising the consumer to seek fuller explanation from the prescribing physician or other doctor of any such information of concern to the consumer." (p.70)

THE ACCEPTABLE DAILY INTAKE (ADI) ISSUE

The FDA has increased the maximum recommended ADI of aspartame from 20 mg/kg to 50 mg/kg body weight. (One kg is 2.2 pounds.) Accordingly, a 130-pound adult would have to ingest four to five liters of an aspartame-containing drink to exceed the new ADI!

How did the FDA arrive at this number? Dr. Frank Young, the present FDA Commissioner, told a Senate hearing on November 3, 1987 that the ADI—in the absence of human data—represents a projection of *animal* studies based on *lifetime* intake. He defined it as

". . . an estimate of the amount of the food additive, expressed on a body weight basis, that can be adjusted daily *over a lifetime* without appreciable health risk. It focuses on continued exposure *over a lifetime* and not on the exposure that could occur on any given day." (Italics supplied)

The FDA has been taken to task for this arbitrary decision.

- Senator Metzenbaum pointed out in Senate hearings held on May 7, 1985 and August 1, 1985 that such action represented an exception to the usual 100-fold safety factor used by the FDA as a guideline for regulated food additives.
- Dr. M. Adrian Gross, a former FDA investigator-scientist, challenged the licensing of aspartame. He stated in the August 1, 1985 Senate hearing:

  "I would view the Acceptable Daily Intake (ADI) set by the FDA for aspartame (50 mg/kg body weight/ day) *as totally unwarranted and extremely high in that it can be associated with completely unacceptable risks as far as the induction of such (brain) tumors is concerned.* It is clear that risks of this magnitude for what the FDA regards as a 'safe' level of exposure to aspartame represent an outright calamity or disaster." (Italics supplied)

- Dr. W.M. Pardridge (1987a) pointed out that the 10 mg/kg aspartame intake regarded by Levy and Waisbren (1987) as "large amounts" could be ingested by a 50-pound child as a *single* 12-ounce can of such a carbonated drink.

My calculations indicate that the maximum daily intake tolerated by most reactors to aspartame products ranges from 10 to 18.3 mg/kg. Certain patients actually could "titrate" the amount on the basis of *predictable* recurrence of itching, rash, headache, mental confusion, depression, or visual changes once the threshold amount had been exceeded. On the other hand, it is obvious that persons who develop severe reactions to small amounts of an aspartame product (for example, grand mal convulsions after chewing a stick of gum) did not even begin to approach the "official" ADI.

A 31-year-old woman experienced her first convulsion 12 hours after ingesting two liters of an aspartame-sweetened soft drink. This amounted to 18.3 mg/kg of her body weight.

A 35-year-old woman developed diarrhea after ingesting a diet drink, as did two close relatives. She noted

"Tolerance levels seem to vary. Only one or two cans of the soft drink affect my mother immediately. For me, it took more. If I had one glassful during the day—no problem. But at several throughout the day or over two days, the problem will begin."

A 34-year-old woman predictably developed a severe itch and rash after drinking four 8-ounce glasses of a diet cola in one day. Lesser amounts (as four glasses a week) were tolerated. The estimated adverse biologic threshold for this 50 kg woman was calculated to be 11 mg/kg.

POTENTIAL "DISINFORMATION"

The issue of aspartame safety has been clouded by (1) pronouncements of reassurance from government agencies and industry-funded professional associations, (2) arguments based on "controlled double-blind studies" that I regard as flawed for reasons repeatedly mentioned in prior sections, (3) the criticism of "anecdotal" case reports invoking reactions to aspartame products, and (4) sophisticated public relations (PR) campaigns underwritten by vested interests. The latter place great emphasis upon the fact that aspartame contains "the building blocks of protein". Some examples of these influences are cited below.

- The administration of dry aspartame in capsules taken with cool water for double-blind studies minimizes both blood phenylalanine levels, and the concentrations of aspartame breakdown products resulting from heat and prolonged storage.
- The FDA Commissioner who approved aspartame stated

  "Few compounds have withstood such detailed testing and repeated close scrutiny, and the process through which aspartame has gone should provide the public with additional confidence of its safety" (July 1981, 46 FR 38285).

- The Chairman of the Professional Advisory Board of the

Epilepsy Institute (*not* the Epilepsy Foundation) distributed this letter on March 18, 1986:

"Recent publicity on aspartame and seizures may have prompted calls from concerned patients. As an organization devoted to people suffering from seizure-related problems, we at the Epilepsy Institute investigated this allegation and found aspartame to be safe for people with epilepsy."

- The FDA makes repeated reference to the "anecdotal" nature of symptoms in a report by the Centers for Disease Control (1984) concerning its "passive surveillance" of 517 complainants. Two-thirds experienced headache, dizziness, and mood alteration, and other neurologic or behavioral complaints. Others complained of gastrointestinal symptoms (24%), allergic skin reactions (15%), and menstrual disorders (6%). Furthermore, the CDC's in-depth analysis of 231 persons from this group revealed "13% reported that symptoms recurred after rechallenge with more than one product, and another 15% reported that symptoms recurred on second use of the same product."

- To my knowledge, no formal public hearings were held in 1983 prior to the licensing of aspartame for soft drinks. Even members of the 1980 Public Board of Inquiry (PBOI) were not notified of such pending action. (The PBOI had been convened by the FDA to evaluate the issue of aspartame-related brain tumors. It recommended that aspartame *not* be approved.) Professor Walle Nauta of the Massachusetts Institute of Technology, who chaired the PBOI, indicated he was under the clear impression that aspartame would be *excluded* from soft drinks (*Congressional Record-Senate* May 7, 1985, p. S5503).

- Producers of aspartame products and their representatives point to its approval by dozens of regulatory agencies in other countries, and by the World Health Organization, whenever the safety of such products is challenged. As far as I have been able to determine from a review of the world literature available to me, however, *the vast majority of these agencies accepted company-sponsored research without ever having done independent confirmatory studies—especially human trials.*

Concerning the general issue of "controlled" studies, few would disagree with their importance—including the author (Roberts 1988e). This is not to say, however, that all other evidence should be ignored. Dr. Denis Mehigan (1981) offered this commentary and suggestion about the status of controlled studies.

"Much traditional medical dogma has fallen victim to the all-conquering controlled trial. Only the foolhardy still question the value of such studies. However, few physicians have both the time and expertise to scrutinize each trial, carefully assess its correctness, and determine how it should affect their clinical behavior, if at all. Most depend on the judgment expressed in editorials or reviews by apparently impartial observers. Some accept only results that are reported in a select group of journals. The extent to which the latter shortcuts may be relied on is, however, not clear, and physicians require some guidelines on this matter. I am proposing a controlled trial of the efficacy of believing in the results of controlled trials."★

QUESTIONS CONCERNING ASPARTAME RESEARCH

Corporate-Sponsored Research: General Considerations

A grateful American public heretofore accepted the integrity of biochemical research and scientific regulatory bodies as an article of faith. This attitude has eroded in the wake of repeated reports of "sciencegate." (A pun asserts: "Mice may become transformed into rats when the cat isn't watching the laboratory.")

- Drs. David L. Schiedermayer and Marc Siegler (1986) wrote a significant article entitled, "Believing What You Read: Responsibilities of Medical Authors and Editors," due to concern over the increasing number of retractions in medical journals because of scientific inaccuracy.
- Dr. Joseph D. Sapira (1988) bemoaned the dying "spirit of clinical investigation" in light of the apparent dissociation between biomedical investigators and the "healing guild of

★ ©1981 *New England Journal of Medicine*. Reproduced with permission.

medicine," especially as university hospitals aggressively compete for dollars.

- There are increasing references in the medical and scientific literature to "the medical-industrial complex" and "the corporate compromise."

Subsidized "entrepreneurial science" aimed primarily at promoting specific commercial products (see below) demands closer scrutiny. Many are shocked to learn that payment for research services and testing may be made by contract (as in any other business) rather than by grants to scholars. In the high-stakes realm of food additives that were marketed without adequate testing, Ralph Nader (1971) noted the ability of large corporations to mesmerize both academic analysts and big governmental bureaucracies relative to "avoiding the arduous task of testing them against empirical reality."

Other Economic Issues

Some examples of economic factors that have contributed to erosion of the coveted aura of respect for biomedical research are listed.

- *Corporate and entrepreneurial exploitation of research for potentially vast profits.* The Lancet editorialized on "Pharmaceutical Funds for Clinical Research: A Mixed Blessing" (January 31, 1987, pp. 257–258):

  "It is a general truth that we get what we pay for in medical research as elsewhere . . . The major drive will be towards the commercial project, and the emphasis on market-oriented research may be increased by the attitudes of universities themselves. Pressed for general funds, they demand high percentage overhead payments from commercially funded work.★

- *The advantage of corporate funding in getting research manuscripts accepted and published by "peer-reviewed" journals* (see below).
- *The preferential diversion of limited research monies to "friendly" investigators by "nonprofit" organizations that are in fact underwrit-*

★ ©1987 *The Lancet*. Reproduced with permission.

ten by big industry. *The Lancet* referred to such corporate-sponsored research in academic centers as "privatized pharmacology" (December 19, 1987, p.1439.)

Criticism of Aspartame Research by Others

There have been repeated claims that research demonstrating the safety of aspartame represents a new standard for excellence in studying food additives. By contrast, some investigators (other than the writer), elected public officials, and journalists with impeccable credentials have expressed criticism over the deficiencies in aspartame research noted in this and preceding chapters.

- Dr. Alexander Schmidt, FDA Commissioner from 1972 to 1976, was quoted as referring to the testing of aspartame as "incredibly sloppy science" (Graves 1984).

- Senator Howard Metzenbaum (1985) expressed this reservation about the pre-licensing testing of aspartame at a Senate hearing held on August 1, 1985: "The Board (Public Board of Inquiry) has not been presented with proof of a reasonable certainty that aspartame is safe for use as a food additive under its intended conditions of use." He then wrote Senator Strom Thurmond, Chairman of the Senate Judiciary Committee, on February 3, 1986:

  "Key seizure test on NutraSweet was never investigated by the grand jury. During a Searle sponsored monkey test, all the animals receiving medium or high doses of NutraSweet experienced grand mal seizures (Doc #28). Searle never performed autopsies. The FDA said Searle made at least four false statements and entries in the report of the study (Doc #1). Though the FDA later claimed it did not rely on the study to prove safety, the seizures were never explained. Failure to account for these seizures is of particular significance given current concerns expressed in the scientific community on precisely this issue."

- Dr. M. Adrian Gross, a respected FDA scientist, was given the task of investigating the quality of experimental studies on aspartame carried out by, or for, the G. D. Searle & Co. prior to its licensure. He submitted a summary of dozens of "serious deficiencies" pertaining to the design, execution, analysis of the manufacturer's evaluation of aspartame to Senator Metzenbaum on October 30, 1987. A few are listed.

- The poor quality of material prepared for microscopic examination of tissues
- The lack of training by personnel making observations in teratogenicity (birth defects) studies
- "The false values presented"
- The excision of tumor masses before death
- The "substantial" loss in pathology information due to technical errors
- Improper departure from protocol specifications
- The failure to report all tissue masses to the FDA, especially brain tumors
- The improper use of pesticides in areas where studies were conducted
- Questionable record-keeping and entries
- The substitution of some animals
- Intercurrent disease among test animals to whom drugs were administered

- The Government Accounting Office (GAO), reporting on its two-year investigation of aspartame during July 1987, stated that more than half of the scientists surveyed were concerned about the neurological reactions and other potential adverse effects of aspartame products in children. Moreover, 40 percent called for further research, 32 percent sought new warnings, and 15 percent suggested a total ban!

Other Shortcomings

Physicians and consumers who seek information from producers and governmental agencies about reactions to aspartame products are offered scores of reports published in scientific journals that seemingly reinforce the issue of their safety. I have previously discussed the major shortcomings of these studies. Let me summarize some of my objections:

- The failure of investigators to detect, or to report, tumors of the brain, uterus and ovary to the FDA.
- The failure of the senior investigator of a "negative" double-blind study on behavioral reactions (presented at an international conference) to know, or to recall, the details of how the aspartame given to children being tested was constituted and administered when I asked for such information.
- The failure of double-blind studies on patients with alleged aspartame-induced headaches and seizures to use the *same* incriminated commercial products (easily obtained in any store) rather than aspartame-containing capsules.
- The failure to test *male* mice for tumors of the urinary bladder following aspartame implantation.
- The failure of an investigator to even respond to my inquiry about his published statistical data, from which he had concluded that aspartame does not adversely affect non-insulin-dependent diabetics.
- The failure of FDA administrators and attorneys to investigate animal studies (particularly seizures in monkeys) before the five-year statute of limitations for prosecution expired—specifically, on October 10, 1977 and December 8, 1977.
- The failure of the FDA to act on a major concern expressed in correspondence (dated January 10, 1977) from Richard A. Merrill, its own Chief Counsel. He requested a grand jury investigation of the manufacturer for its alleged "concealing material facts and making false statements in reports that the animal studies conducted to establish the safety of the drug Aldactone and the food additive Aspartame." Mr. Merrill prophetically projected

"The FDA must receive the truth, not psychological warfare. *To emphasize the importance of safety data on aspartame, we note that if ultimately approved for marketing, this sweetening agent can reasonably be expected to be part of the daily diet of every American.*" (Italics supplied)

- The failure to challenge the manufacturer's contract with Universities Associated for Research and Education in Pathology (UAREP). This private group was engaged to determine the factual accuracy of prior aspartame studies—*but* with the stipulation that UAREP "shall not express an opinion" regarding either the design or safety significance of these studies, nor make recommendations about the safety of aspartame for human use! Dr. M. Adrian Gross (see above) also challenged the credentials of UAREP relative to its ability to assess prior aspartame studies.

- The failure of the then-FDA Commissioner, who had held this office only several months, to heed the strong dissenting opinions of in-house FDA scientists concerning aspartame licensure.
- The failure of the NIH to renew research grants in basic pharmacology concerning the metabolism of aspartame, notwithstanding prior investigations that gave promise of major insights. One involved a potentially simple blood test to detect a deficiency or genetic variant of an enzyme responsible for the intestinal breakdown of aspartylphenylalanine (Chapter 8).

The Influence of Corporate Sponsorship

Reference was made earlier to the extent of self-serving research with the expanding "industrialization of academic medicine" (Freedman 1986). As an unfortunate corollary, it now becomes necessary to scrutinize the results of such research— including aspartame studies—for undue partiality toward corporate sponsors.

- Researchers conducting tests on aspartame have received corporate largesse in various forms, such as major donations to their institutions. The laboratory of one aspartame researcher allegedly received $1.3 million.
- The services of respected physicians who advocate both the use and safety of aspartame products in managing diabetes and obesity are valuable. Their corporate value is enhanced when they function as (1) "experts" at Congressional hearings, (2) representatives of medical and scientific societies on nationwide television programs that focus on aspartame reactions, and (3) reviewing consultants for "peer-reviewed" medical journals to which manuscripts or letters challenging the safety of aspartame are submitted.

As noted, corporate interests have immediate access to prestigious scientists—both in-house and handsomely paid consultants—who can (1) counter *any* article or presentation that casts suspicion on the safety of aspartame products, or (2) serve as expert defense witnesses for depositions or at court in cases of alleged product liability. At the very least, such rebuttals and delaying tactics enable the continued generation of great profits until the FDA or some other regulatory body acting in the public interest might finally decide that an imminent public health hazard indeed does exist.

PROFESSIONAL OBSTACLES

Any professional who has tried to obtain *detailed* information about aspartame reactions and toxicity from the manufacturer, contracted researchers, or the FDA can vouch that the task is not easy. The same applies to getting medical reports about reactions to aspartame products—or even a letter to some editor—published.

Yet another thorny issue confronts practicing doctors, dietitians and consumers. It can be phrased as this question: "Who *does* one believe when there is heated controversy about safety in the case of such popular products already approved by the FDA?" FDA Commissioner Young aptly stated at the November 3, 1987 Senate hearing on aspartame: "You will be dazzled and at times confused by a wide variety of scientific presentations."

The attitude generated among physicians concerning the rarity of reactions to aspartame-containing products and the "anecdotal" nature of published reports accounts in part for their failure, or unwillingness, to identify them. Beckman and Frankel (1984) emphasized the frequency with which physicians prematurely interrupt opening statements by patients ... 69% in their experience. This reflexive response clearly could result from the disbelief of doctors about relevant information offered up-front by patients who suggest that aspartame products might be playing a major role in their problems.

The potential multiplicity of corporate and bureaucratic obstacles confronting individuals or small firms who challenge large companies has been described by Morton Mintz and Jerry S. Cohen in *America, Inc.: Who Owns and Operates the United States* (Dell Publishing Company, New York, 1971). In his introduction, Ralph Nader emphasized some of the "relentless" strategies used ... including friends in governmental bodies responsible for regulating "harmful food additives."

Obstacles by Medical Journals and Publishers

I can personally vouch for the *enormous* difficulty in getting published articles concerning reactions to aspartame products. Many "peer-reviewed" journals refused to publish *one* of more than a dozen short "letters to the editor" pointing out flaws in protocols, or detailing the scientific criticism of reports on aspartame they had published.

> I submitted a short manuscript to *Neurology* entitled, "Aspartame-Associated Confusion and Memory Loss: A Possible Human Model for Early Alzheimer's Disease." It was based on data involving 157 patients. This manuscript was *promptly* returned with this one-sentence letter from the editor: "The conclusions in your paper were not derived by a replicable objective controlled study and therefore we cannot consider your paper for publication in *Neurology*." (Considering the increasing magnitude of Alzheimer's disease, and the relevance of my observations to newer biochemical findings and avenues for research [Chapter 28], the readers of this publication were denied access to an "original" and potentially useful perspective.)

The extreme reluctance of journal editors to publish reports on controversial subjects, such as reactions to aspartame products, has ominous overtones. An opinion by the Supreme Court (*Red Line Broadcasting Co., Inc. v Federal Communications Commission,* 89 S. Ct. 1794, 1969) has special relevance to magazine and book publishers owned by corporate conglomerates that also have vested interests in foods, drugs and chemicals.

> "It is the purpose of the First Amendment to preserve an uninhibited marketplace of ideas in which the truth will ultimately prevail, rather than to countenance monopolization of that market, whether it be by the Government itself or a private licensee."

Admittedly, some editors anguish over the selection or rejection of controversial manuscripts. A commentary by Dr. James L. Mills

(1987) in the *Journal of the American Medical Association* (*JAMA*) was entitled, "Reporting Provocative Results: Can We Publish 'Hot' Papers Without Getting Burned?" He addressed the increasing difficulty confronting authors and editors alike relative to publishing "provocative results" and "hot papers" in "our current litigious and socially volatile climate." How, then, do researchers communicate unpopular findings? Their options are generally limited to "burying" the findings in a small-circulation journal (such as the bulletin of a county medical society), reporting the results as a letter to the editor, or (unfortunately, most often) discarding the project.

Journals which base acceptance or rejection of articles in large part on the reviews of "established" workers in the field command by far the greatest attention. These peer-review journals, however, may not be entirely free of bias.

- Dr. Lawrence Altman (1988) expressed concern over the extraordinary influence exerted by the prestigious *New England Journal of Medicine* relative to the flow of information (or lack thereof) on medical research, the treatment of patients, and government policy. A respected Princeton health economist was quoted as regarding such manipulation of information as ". . . all in the name of peer review. But it really is in the name of profit."
- I disagree with editors of the *JAMA* who contend that "all the important advancements in science are published in peer-reviewed journals" (Rennie 1989). Novel observations and concepts by observant clinicians practicing in the trenches of medicine, however, might not be able to provide "statistically significant" support for "passing peer review." Even these editors recognized that (1) interested editors and those with self-serving interests can get a publication "fast-tracked," and (2) manuscripts may be rejected when editors select "unusually rigorous, even hostile, reviewers."

Reassuring Statements

Dozens of statements by "authorities" have been offered to reassure professionals about the safety of aspartame products.

- A book review in the *New England Journal of Medicine* stated that non-compulsive aspartame intake has "no sinister effects" (Holliday 1986).

- A laboratory investigator advised against wasting further effort relative to aspartame reactions by "being obsessive about something that isn't really there" (Fernstrom 1987).
- Dr. F. Xavier Pi-Sunyer (1988), representing the American Diabetes Association's Board of Directors, stated at a Senate hearing on aspartame that he objected to this session on the grounds it might make the public needlessly apprehensive.

"We have no indication from these professionals (physicians treating diabetic patients) that there are significant problems with the use of aspartame, or, for that matter, that there is any pattern of complaints regarding this product. The use and safety of aspartame simply has not been an issue among these health professionals, as far as we can determine . . . We do not believe, however, further studies are needed to duplicate current knowledge . . . In conclusion, ADA supports the continued availability of aspartame. Our organization believes the product has been shown to be safe."

The Absence of Dissent

There is still relatively little information in the medical and scientific literature concerning the frequency and severity of reactions to aspartame products. Several phenomena probably have contributed to this state of affairs.

- *Intimidation by the prospect of costly legal battles that may be initiated by manufacturers and producers of aspartame products who have virtually unlimited funds to engage in, and prolong, such litigation.* One aspartame reactor was told by a university center physician that the serious neurologic and ocular reactions "possibly have been caused by aspartame . . . but I couldn't put it in writing for fear of a lawsuit." Having experienced the hurdles of raising two chemically sensitive children, Ken and Jan Nolley (1987) commented that " . . . the current unrelenting threat of malpractice litigation produces pressure to avoid diagnoses not clearly and unambiguously validated by the tests the physicians have at their disposal."

- *The soft-pedaling of errors in corporate-sponsored research by reviewing scientists who fear product libel suits (Science 235: 422–423, 1987).* In the case of aspartame products, this could apply to flawed "negative" double-blind studies (for reasons cited throughout this book).

- *The ever-present potential influence of "medical politics" on the peer review process, especially from referees having vested interests.* Consultants for this billion-dollar industry may be influenced by financial considerations. Dr. Richard Wurtman told a Senate hearing on November 3, 1987 that a number of prestigious investigators paid as private consultants to this industry chose *not* to attend the recent highly pertinent international conference he had chaired (1987c).

- *The enormous revenues reaped by medical journals and lay periodicals from the advertising of aspartame products.* A personal encounter illustrates this issue.

The publisher of a household-name magazine made an *urgent* night call to me at my home, requesting that I submit an article on aspartame reactions that "positively" would be printed. I even was asked to send the manuscript "collect" by an overnight express service. There was another reason: *two* children of this publisher had suffered severe reactions to aspartame products. I received a call one week after sending the manuscript. After beating around the bush for several minutes, this publisher confided, "When I showed it to my advertising people, they hit the ceiling because so much income could be lost if companies making or using aspartame pulled their ads!"

The likelihood of subsequent class-action litigation for failure to provide post-sale warnings of products subsequently determined to cause injury is real. The issue was raised in *Yuhas v. Mudge* (1974), as to whether a publisher of a magazine can be held responsible when a defective product advertised in the magazine resulted in injury . . . even though he did not manufacture, distribute, endorsee or sell the product. While the trial judge determined that no actionable duty existed, this issue may be again raised if publishers are presented with scientific and epidemiologic evidence for the probability of injury to an advertised product, and then continue to advertise the product.

Inability to Address Medical Associations

The difficulty encountered in presenting information on aspartame reactions to medical associations—or even having an abstract published in their proceedings—is noteworthy. Much of this problem could relate to the positions taken by major agencies (e.g., the FDA) and associations (e.g., the American Diabetes Association).

- I submitted an abstract for presenting a report to the 1987 annual scientific program of the American Diabetes Association. Entitled, "Complications Associated With Aspartame In Diabetics," it summarized observations involving 58 diabetic aspartame reactors (Chapter 25). This "paper" was rejected for presentation, as was publication of the abstract. The Chairman of the Committee on Medical and Scientific Programs replied that his committee

  ". . . does not find it suitable for publication in the program supplement. We receive a large number of abstracts and all are carefully reviewed by the members of the Committee and ad hoc reviewers. It is our policy, however, to publish only those which are judged to be of clear scientific merit."

- I submitted another abstract for the 1988 annual scientific program of the American Academy of Neurology entitled, "Aspartame-Associated Epilepsy." It reviewed data on 250 such patients—101 cases from my series, and 149 cases reported to the FDA. The letter of refusal contained this ironic twist: "Please submit for future meetings."

LEGISLATIVE OBSTACLES

The reader must bear in mind an important caveat: *the public cannot expect aspartame producers, the "health food" industry, or the "wellness press" to dampen their enthusiastic promotion of highly profitable products—even in the face of serious challenges concerning safety—unless forced to do so by restrictive legislation.* Considering their enormous biopolitical power in an era of "deregulation," such timely action is not likely to occur without the stimulus of a thalidomide-like disaster.

Yet, federal investigators and legislators dare not ignore this matter until "absolute" proof is forthcoming. By contrast, I consider the ongoing labeling of saccharin products as potentially causing cancer to be irresponsible (Chapter 34), and unjustly partisan to corporate competitors.

A 31-year-old woman wrote

"I believe it (aspartame) was pushed through as an alternative to saccharin without being investigated. 'BIG BUCKS' were to be made by many if aspartame became a success, which it did, but its continued use could be devastating."

Senator Howard Metzenbaum

Senator Metzenbaum (1985) correctly asserted that *the burden of proof concerning safety of the aspartame products is the responsibility of manufacturers and producers . . . NOT government.* He proposed the Aspartame Safety Act of 1985 (S. 1557). It would have mandated independent studies conducted under the auspices of the National Institutes of Health, and appropriate informative labeling. The Senate defeated his proposal by a 68–27 vote.

Gregory Gordon (1988), an investigative reporter for United Press International, subsequently listed the sizable recorded contributions to several influential senators from corporate officials and their families not long after they cast the negative votes favorable to their firm, which manufactured aspartame and products containing it.

The disinclination of the United States Senate to approve this proposed Act contrasts with its overkill legislation that previously banned the sweetener cyclamate.

Senator Metzenbaum later sought full subpoena power to investigate allegations about the concealment of "material facts" and the making of "false statements" concerning the licensing of aspartame. Once again, he was unsuccessful.

Suggestions

In the face of *thousands* of volunteered complaints by aspartame consumers, Congress ought to demand autonomous *"second opinion" research on aspartame—including the replication of ALL reported studies of its effects on humans and animals.* Several additional suggestions are offered.

- The investigators should be corporate-neutral, and function under the auspices of the National Institutes of Health.
- Aspartame products purchased and prepared under real-life circumstances should be employed (Roberts 1988e).

OBSTACLES CREATED BY THE FDA

The Food and Drug Administration (FDA) has come under heavy fire relative to (1) its blanket approval of aspartame products for public consumption, and (2) failure to enforce its own standards . . . specifically as they relate to aspartame. Indeed, the request by the FDA's chief attorney for a grand jury investigation of research on aspartame apparently was the second in this Agency's history.

- Senator Metzenbaum asked the FDA at a Senate hearing concerning aspartame, held on November 3, 1987:

  "In the history of the FDA, has there ever been any other case in which the scientific data submitted to the FDA caused the FDA to be so concerned about the representations made to it that it was sent to a U.S. attorney for presentment to a grand jury for indictment—any other cause? . . . it is my understanding—I think this occurred in 1977—that prior to that time there had never been a case in which the submissions of a company to the FDA had provided a basis to submit their file, their record, to the U.S. Attorney for submission to a grand jury, and that since then there has certainly been none. Is that correct?"

- Thomas Scarlett, Chief Counsel for FDA, replied to the Senator on December 22, 1987: "There was at least one case referred to the FDA to the Department of Justice for the consideration of criminal charges because of irregularities in the reporting of test data. The case involved the drug MER-29 and was referred sometime in the 1960s." (I had described the occurrence of hypothyroidism following the administration of this cholesterol-lowering drug [Roberts 1965d], bearing the generic name triparanol.)

Senator Metzenbaum criticized the FDA as ". . . more of a handmaiden of the food and chemical industry than it is a defender of the health and safety of American consumers" during the August 1, 1985 Senate hearing. He subsequently noted that *ten* ranking FDA or federal officials involved with the investigation and regulation of aspartame had left government service for employment by that industry! They included the FDA Commissioner who initially approved its licensure, and former chiefs or acting chiefs of the Bureau of Foods.

- During June 1983, one month before the FDA approved the addition of aspartame to soft drinks, the FDA Commissioner replied to questions about the safety of aspartame breakdown products:

  "The agency also concludes that the allegation that all possible reaction products have not been tested for safety is not a tenable issue in terms of regulatory food additive safety evaluations. Such a requirement for the demonstration of safety would necessitate an unlimited amount of experimental data" (48 RF 142 P 31383).

- The *prompt* precipitation of severe neurological and medical problems within minutes or hours after the ingestion of aspartame products, including grand mal convulsions (Chapter 9), casts doubt upon the FDA assertion that "the agency does not regard the possible consumption of aspartame in a single large dose as posing any safety problem whatever" (*Federal Register* February 22, 1984, p. 6678.)
- Pharmaceutical companies are required to mail "Dear Doctor" letters to physicians when there have been repeated reports of side effects to any new drug. By contrast, such

vigilant post-marketing surveillance is *not* required for approved food additives.

The *predictable* sequences of complaints encountered by aspartame reactors have been elaborated throughout this book. Yet, Dr. Frank E. Young, FDA Commissioner, told a Senate hearing on November 3, 1987.

"As of October 23, 1987, the ARMS [Adverse Reaction Monitoring System] has evaluated 3,511 of the 3,679 aspartame consumer complaints received by the FDA. There is still no consistent pattern of symptoms reported that can be attributed to the use of aspartame."

"Moreover, because the reports are most frequently anecdotal in nature, it is not possible, without other data such as medical records, to eliminate factors other than aspartame consumption as possible causes of the reported effects." (I had *personally* handed my medical records on a young nurse present at this hearing to an FDA investigator. Her first epileptic seizure occurred within one week after beginning to use aspartame beverages. She was not allowed to testify.)

The ingrained attitude of the FDA relative to the subject of aspartame reactions is clearly evidenced by the following exchange at a hearing on aspartame by the Committee on Labor and Human Resources of the United States Senate held on November 3, 1987.

Senator Orrin Hatch: "Do you persist in your opinion that the 3,000 adverse reaction reports you have received are of no clinical significance?"

Frank Young, M.D. (FDA Commissioner): "Yes, I do, and this is a relatively low incidence of adverse reactions compared to the large number of individuals who are using the product."

Senator Hatch: "Has the Center noticed an increase in adverse reaction reports associated with the publicity on aspartame?"

Commissioner Young: "Yes, we have. We usually see a peak of adverse reactions with publicity."

Other related criticisms of the FDA are germane.

- Dr. M. Jacqueline Verrett, a former FDA biochemist and toxicologist, told the Senate hearing on November 3, 1987 that she had found serious departures from standards during her analysis of the original studies performed by the manufacturer in the early 1970s. Uterine polyps, altered blood cholesterol and other changes were discarded as "minor" findings. She testified

  "It is unthinkable that any reputable toxicologist giving a complete objective evaluation of this data resulting from such a study could conclude anything other than that the study was uninterpretable and worthless and should be repeated. This is especially important for an additive such as aspartame, which, as we have heard already today is intended for and is now being used in such a widespread and uncontrolled fashion."

- FDA Commissioner Young told this same Senate hearing that he regarded the monies spent on aspartame research as grossly disproportionate. Paradoxically, he also referred to the FDA's "poor facilities" for such evaluation.
- Foods claimed to prevent disease are considered "drugs"— and therefore subject to pre-market verification for both safety and effectiveness. On the other hand, the FDA has consistently tolerated the implied health claims for "low-calorie" and "reduced-calorie" products containing aspartame. Particular attention is directed to the problems encountered with their use for weight loss and diabetes management, discussed in previous chapters.

I have sought other explanations for shortcomings of the FDA. One was raised after being asked by a key professional to supply the *actual* reprints of several references I had listed in my public comment concerning "fake fat" (Roberts 1989a). To my astonishment, I was informed that (1) the FDA staff has limited access to reference material, and (2) literature searches are not readily available due to limited funds because they must be specifically contracted. This obviously prevents those entrusted with protecting the consuming public from doing the required research. If a classic example of pennywise-and-pound-foolish was needed, this policy would qualify.

JUDICIAL OBSTACLES

Consumer advocates were placed at a great disadvantage by the refusal of the Supreme Court to investigate approval of aspartame for public use after a prior ruling by the U.S. Circuit Court of Appeals for the District of Columbia (*The New York Times* April 22, 1986). Specifically, the Court would not hear claims challenging the safety of aspartame-containing soft drinks by the Community Nutrition Institute and other consumer groups.

A subsequent major ruling by the Supreme Court in *Berkovitz v. United States* (1988), discussed in Chapter 34, allowed for a federal agency to be sued if there had been negligent violation of governmental regulatory policies.

CHAPTER THIRTY-THREE

Attitudes of Consumers and Patients

Unchecked giantism—be it political, religious or economic—has never been compatible with individualism.
— SENATOR PHILIP HART
(1968)

CONSUMERS WITH APPARENT REACTIONS to aspartame-containing products understandably have expressed frustration and anger as a result of their costly experiences and the futile search for accurate relevant information. This bitterness has been directed toward the FDA, the Congress, the producers of aspartame products, and the medical profession.

A 61-year-old woman had epilepsy. She remained free from seizures for 12 years until the attacks recurred after taking an aspartame product. She wrote the manufacturer, "But, is it not better to be honest about the possible bad effects of aspartame, or must you change only after the company is slapped with a huge lawsuit?"

Some representative expressions I have received are reproduced in the following pages. As noted in the Preface, these remarks were volunteered as part of a research effort, and do not necessarily reflect the attitude of the author or publisher.

"Do you think that Congress should do anything about regulating the use of aspartame?

This question in the survey questionnaire elicited vehement comments from many aspartame reactors.

"Ban it!"

"Take it off the market!"

"Stop its sale!"

"Remove it from the market until they can prove it harmless!"

"To put it in children's vitamins is unforgivable!"

"Congress must forbid it on the market!"

Scores of accompanying explanations pleaded for the capping of aspartame consumption pending further impartial and corporate-neutral testing of its safety.

- A legal secretary advised:

  "Restrict it! There is too much on the shelves. It's an inconvenience for me to encounter aspartame in every diet food (almost every). I know it makes me ill, so I avoid it. However, it is in so many things and children ingest so much of it. I wonder how many of them suffer and are not able to discern what is upsetting them. I do not believe it should be so universally used."

- A married couple and their daughter—all aspartame reactors—wrote:

  "Stop using it! So many people have problems. You don't know what causes it until you hear or see or talk with someone who says something about a problem with aspartame. STOP IT, PLEASE."

- A 26-year-old woman with aspartame-associated seizures stated:

  "It's frightening to think this sweetener was put on the market when not enough was known about it."

- A 71-year-old woman developed headaches, drowsiness, unsteadiness, "loss of energy," memory impairment, unex-

plained chest pains, heartburn, nausea and itching while
.taking one packet of an aspartame tabletop sweetener daily.
Her complaints abated within three days after abstinence
from this product. She wrote:

"If aspartame affects me so drastically, there must be thousands of other people who are affected and don't expect the
source to be aspartame . . . I think aspartame is a very dangerous product and should be removed from the market."

- A 41-year-old secretary who suffered aspartame-associated
  seizures and striking personality changes noted:

  "After what it has done to my body and my entire life
  . . . Why should anyone have to gamble to see if they happen
  to be one of the sensitive ones?"

- A 42-year-old registered nurse with severe aspartame-associated depression anguished:

  "My worry is that they'll possibly just put a warning label
  on the product and not take it off the market completely."

OTHER EXPRESSIONS OF CONSUMERS' ATTITUDES

Many criticisms were leveled at the licensing and marketing of
aspartame products.

- A secretary with headaches, personality changes, and multiple
  grand mal seizures attributed to aspartame wrote:

  "Naturally, I am angry now. I have been through a terrible
  physical shock. My husband has been put through mental
  anguish—he still has nightmares about that night (of my first
  convulsion). And I am afraid to be alone for fear I might have
  another attack and there would be no one to help me. I know
  I'll eventually get over that, but it's a very frightening thing."

- A 42-year-old teacher developed dizziness, lightheadedness,
  exhaustion, confusion, slurred speech and a rash while using

aspartame products. Her complaints disappeared one week after avoiding them, and recurred within three days on rechallenge. She stated:

> "I no longer eat foods with aspartame in them. I read labels and ask questions. This includes foods, sodas (I now drink water), gum, candy, etc. Within eight days of being off aspartame on two separate occasions, all symptoms disappeared. Playing golf became a pleasure. Within weeks, I was back to riding my bike at least 12 miles a day.
>
> "Two years ago, I attended an ice-skating championship sponsored by an aspartame product. I wore a T shirt that said in bold letters, 'Aspartame is dangerous to your health,' and passed out flyers."

This tirade of "righteous indignation" occasionally had a sprinkling of humor. One teacher with severe reactions to aspartame-sweetened tea quipped, "What we need is another Boston Tea Party."

Ridiculed Complaints

Aspartame reactors generally become infuriated when they are told that (1) they have a rare idiosyncracy, (2) such "anecdotal" reports are meaningless, and (3) they were self-diagnosing themselves as "victims" (*The Miami Herald* February 16, 1986, p. A–18). These repeated encounters served to crystallize their ineffectiveness as individual consumers . . . and the need for legal and political defense (Chapter 34).

CRITICISM OF THE MEDICAL PROFESSION

Aspartame reactors were often unhappy about the evasiveness or perceived apathy of doctors who almost reflexively rejected the notion that aspartame products could in any way be causing or contributing to their problems. The refusal of physicians to pursue this possibility—even after patients personally handed them articles and clippings that seemed directly applicable—was especially frustrating . . . and perceived as "brainwashing."

- A 42-year-old registered nurse attributed her severe anxiety, depression and abnormal behavior to aspartame products. She wrote, "I spoke to three doctors about my symptoms being caused by aspartame, and they all just about laughed in my face. How can you work with this?"

- The wife of a food scientist developed tremors, a faint feeling, severe weakness in the lower limbs (once forcing her to lie on the floor while shopping), headaches, depression, intolerance to noise, and behavioral changes. She had been using considerable amounts of aspartame products because of a prior diagnosis of hypoglycemia. Her complaints disappeared within ten days after avoiding them. But she then was scolded by her doctor for having discontinued some of the medications after she became symptom-free!

- A 39-year-old woman stated, "I was amazed by the total personality change. My tests were primarily my own. Doctors think I'm 'nuts' to think a sweetener would be the problem."

Disinterested neurologists were singled out for particular criticism by aspartame reactors with otherwise-unexplained headaches and convulsions.

- A 37-year-old manager experienced non-convulsive "fits," severe headaches, dizziness, memory loss, insomnia, slurred speech, depression, pains in multiple areas, and marked nausea after drinking four liters of aspartame-containing soft drinks daily for several weeks. These complaints improved when he stopped them. After seeing two neurologists, he opted against seeking further consultation because "Doctors here were not aware of any problems with aspartame."

- A 31-year-old female executive experienced three convulsions after drinking four cans of an aspartame cola and three packets of an aspartame tabletop sweetener daily. They did not recur when she avoided aspartame products. She summarized her encounters with neurologists, and her ongoing plight, in the following letter.

"At this time, I can no longer see the neurologist whom I was seeing before. His philosophy is not in tune with my thinking on the subject of aspartame. I would appreciate it if you could recommend a neurologist in the area. I am desperate to find a doctor. I am aware that with my history I should be under a doctor's care."

- A 35-year-old male anesthetist suffered headaches, memory impairment, visual problems and three convulsions while drinking 4–6 diet colas daily. He informed a Senate hearing of his neurologist's reply to the suggestion that aspartame might be contributory: "Wouldn't it be a shame if all that is wrong with you is NutraSweet?" (Taylor 1988).
- A 59-year-old man had his first convulsion while consuming up to eight glasses of aspartame soft drinks and three packets of an aspartame tabletop sweetener daily. He then predictably suffered convulsions after taking aspartame in various unrecognized forms. Other aspartame-associated complaints included headache, dizziness, insomnia, personality changes, mental confusion and memory loss (necessitating retirement from his job.) His wife expressed her displeasure in these terms: "I have asked doctors to report his convulsions to the FDA, and they all refused. They all say the American Medical Association says that aspartame is harmless."

- A 31-year-old registered nurse suffered a grand mal convulsion within hours after drinking two liters of an aspartame-sweetened drink. All of the appropriate studies proved normal. Researching this subject in the medical library, she found three pertinent articles. When she asked the neurologist to consider them in her case, he offered this advice: "Stop reading!"

Related Considerations

Discussing the *plight of the parents* having young patients with chemical sensitivities, Ken and Jan Nolley (1987) asserted: "These dismissals often reveal a disdain for patients' abilities to assess their own symptoms rationally, as well as subtle, unexamined cultural biases." They criticized physicians whose ". . . own interest makes it easiest to deny patient-reported symptoms, and their training reinforces the perceived superiority of their own judgments over patient perceptions."

Consumers also find themselves increasingly subjected to extensive *nutritional self-deception*. It evolves as health authorities, under the mantle of presumed expertise, champion seemingly altruistic nutritional strategies against degenerative diseases. Professor Alfred E. Harper (1988) emphasized that such activities have undermined confidence in the American diet. Moreover, they serve to create "healthy hypochondriacs" by instilling a fear of food (trophophobia)— whether directed to calories, carbohydrates, fats, proteins, or combinations thereof. This perversion already is manifest by undernourished small children from affluent families whose parents consider high-calorie foods rich in nutrients (e.g., dairy products, eggs and meats) to be hazardous. In the process, they encourage "safer" low-calorie and low-cholesterol substances . . . like aspartame products.

CRITICISM OF THE FDA

Suspect reactors to aspartame products have directed their most intense rage against the FDA. One-fourth of those completing the survey questionnaire indicated they had written or called this agency . . . especially after personally deducing that their complaints were directly attributable to one or more aspartame products. Failing to receive a satisfactory answer, or any at all, several asked virtually the same question: "Is the FDA supposed to represent the public or the company?"

- A 48-year-old woman experienced two grand mal convulsions while drinking up to eight cans of aspartame-containing soft drinks daily. There were no further seizures after aspartame was avoided. She then called the FDA to offer her medical records . . . only to encounter total lack of interest.
- A 27-year-old manager with severe aspartame-associated headaches described her fury on viewing a television commercial that suggested aspartame products are natural and healthy. She stated, "My experience taught me a lot. Too bad it is not enough for the FDA to see their potential harm."
- An female aspartame reactor sent me a copy of her long letter to the FDA, dated December 15, 1986. The following excerpts, expressing her outrage, are noteworthy.

"As might be expected, only a small fraction of people who experience problems with aspartame bother to contact the FDA, their congressmen, friends, or anyone else. So don't tell me that you have had only so many complaints. This is the number of people that have bothered to write."

"I would estimate that about 90 percent of my female friends have had problems with aspartame, and have quit using diet drinks and other products."

"This may not represent a health problem to the FDA or any other organization, but it certainly has to these people who have been to doctor after doctor, chiropractors, given up their activities, some have given up their jobs, plus the mental and monetary problems." "What does the FDA consider a health problem anyway?"

"Your letter stated 'This sensitivity to aspartame is presumably similar to the adverse response experienced by some individuals to various other foods and color additives, as well as to certain foods.' Well, somebody presumed wrong. I have never experienced the same problems with anything else in my life, nor have these people with which I'm familiar. It is very strange how you are trying to rationalize this problem, rather than believing the people and removing the problem."

"You haven't received millions of complaints because people don't know that aspartame is causing their health problems. Erroneously, they think that the food and drinks they consume are safe."

"Animals cannot speak and tell you how they feel."

- A 68-year-old woman was seen in consultation for multiple and severe reactions to aspartame-containing products. She wrote me several months later to indicate: "I have been feeling fine, lost weight and even look younger." She stated:

  "Those aspartame attacks were so horrible. A total of at least five hours were shot, as I couldn't function properly."

  "Where is our F.D.A.? The only winners are the impressarios who sit back and count their money . . . Sometimes when I think of the suffering it has caused me and countless others, I could tar and feather each and every one."

- A female with severe hives and menstrual bleeding following the ingestion of aspartame products stated:

"I thought of writing to the company or the FDA, but I figured it would do no good. These people have some conviction that their product is safe beyond a reasonable doubt. I am willing to testify that for me it was very unsafe . . . I am frightened about the growing use of this product in children's foods and so many of our foods. Just think about how much aspartame kids drink during the summer. You can liken this to my drinking lots of iced tea and sodas and what it did to my body . . . I believe there's a national tragedy in the making." (Italics supplied)

Comparable denunciation of the FDA by consumers exists with other products as well. One example is cited.

A person with severe reactions to sulfite food preservatives charged that the FDA made it exceedingly difficult for consumers to verify such responses since "it's too hard to become a statistic" (Korte 1989). Moreover, the proposed changes for reducing sulfite labeling placed them "in an untenable, unnecessary life-or-death situation."

CHAPTER THIRTY-FOUR

Consumer Advocates and Aspartame

*The Government is very keen on amassing
statistics. They collect them, add them, raise them
to the nth power, take the cube root and prepare
wonderful diagrams. But what you must never
forget is that everyone of those figures comes in the
first instance from the village watchman, who just
puts down what he damn pleases.*
— *SIR JOSIAH STAAMP (1929)*

CONSUMER ADVOCATES HAVE TRIED to warn the public—both as individuals and organizations—about reactions to aspartame-containing products. Senator Howard Metzenbaum (1988) informed a Senate hearing that the FDA had received "close to 4,000 consumer complaints, ranging from seizures to headaches to mood alterations."

Some consumer advocates refuse to wait indefinitely for "absolute" proof of a threat to the public health (Chapter 30) with the existing amount of so much incriminating evidence, coupled with bureaucratic stalling.

- In a letter dated February 3, 1986, Senator Metzenbaum wrote Senator Strom Thurmond, Chairman of the Senate Judiciary Committee:

"The average consumer assumes that all safety questions surrounding this sweetener have been resolved long before it found its way onto every grocery shelf in America. A recent investigation undertaken by my office raises serious questions as to whether this is, in fact, the case."

- Commenting on the incomplete nature of all scientific work, Bradford Hill (1965) emphasized that this perception " . . . does not confer upon us a freedom to ignore the knowledge we already have, or to postpone the action that it appears to demand at a given time."

The message ought to be crystal-clear. First, consumer complaints and scientific-medical concern about reactions to aspartame products no longer can be brushed aside by industry or the FDA as "anecdotal" idiosyncrasies. Second, this and other regulatory agencies must avoid a repetition of the aspartame experience in terms of premarketing human trials. This warning is especially timely as licensing will be sought for the extended uses of sweeteners (acesulfame-K, sucralose, alitame) and the introduction of microparticulated protein as "fake fat" (Roberts 1989a).

The Community Nutrition Institute

Rodney E. Leonard of the Community Nutrition Institute, a consumer organization in Washington, D.C., issued this pertinent statement on July 17, 1986. The following excerpts (reproduced with permission) are noteworthy.

"The continued use of aspartame . . . as a food additive endangers the health of too many Americans and should be stopped immediately. The Community Nutrition Institute is petitioning the Commissioner of the Food Drug Administration (FDA) to declare that aspartame is an imminent hazard and to ban the chemical sweetener from use as a food additive.

"The danger is real: aspartame causes harm to some people. It is not a harmless substance.

"We have held this belief for some time. We did not believe in 1981 that FDA had sufficient or credible scientific data to make a finding that aspartame would cause no harm. We said in

1983 that the introduction of aspartame in soft drinks would greatly increase the health risk for no apparent health benefit. By late 1983 and early 1984, individual users began to contact us with complaints of adverse reactions. FDA and (the company) which markets aspartame . . . dismissed those complaints as simply the normal 'placebo' effect that follows the introduction of any new additive. We did not accept that argument . . .

"We are angry. We are disheartened. FDA now has the clinical evidence that people are harmed. It is evidence acquired in the most brutal and heartless way: an experiment has been conducted using the American people as guinea pigs, as test animals . . .

"We believe that if the Commissioner of FDA were truly concerned with health and safety, the facts support a finding of imminent hazard. He has demonstrated a willingness to interpret the law in novel ways. For example, he has introduced the 'de minimis' concept as an interpretation of the Delaney clause in the food safety law. This interpretation allows the use of any carcinogen as a food additive if FDA determines it is present in amounts so small as to be inconsequential. *The Delaney clause specifies that any substance causing cancer in test animals or humans is prohibited from use as a food additive."* (Italics added)

The Aspartame Consumer Safety Network

This national consumer activist group formally requested the FDA on June 27, 1989 to deny two petitions submitted by Foodways National, Inc. and the NutraSweet Company for the addition of aspartame to frozen dairy and nondairy frostings, cheesecakes, fruit and fruit toppings.

This group, headed by Mary Stoddard of Dallas, cited the following facts as reasons for denying both petitions, and for convening a formal hearing thereon.

"1. Aspartame, in actual use by human consumers, has not been proven to be safe. In fact, thousands of consumers have reported Central Nervous System related symptoms to the FDA and consumer groups such as ours. The FDA has in its files numerous forms and petitions that members of our group have provided, detailing various serious medical complaints against aspartame (NutraSweet/Equal). There have been at least 5 reported deaths due to consumption of aspartame, being investigated currently by the FDA and others.

"2. The "double blind" challenge test (placebo & aspartame capsule) is provided to physicians and test results are evaluated by the NutraSweet Co. itself. So far, they have *never* admitted to having anyone who submitted to testing in this manner turn out to have a special sensitivity to aspartame. Yet, in the general population, thousands have quit using it—their problems go away—(and) on rechallenge, the symptoms return. This includes accidental/unknown rechallenge . . .

"3. Pilots & Aspartame. Several pilot's publications have published warnings to pilots and others concerning consumption of aspartame. According to the Editor of Pacific Flyer Aviation News, hundreds of pilots have reported adverse reactions including grand mal seizures and safety-in-flight incidents—risking the lives of pilots and their passengers. Many pilots have lost their medical certification to fly due to their consumption of aspartame. The problem has become so severe that in December, 1988, national leaders of the Aspartame Consumer Safety Network convened a FAA national headquarters in Washington DC for a meeting with Deputy Flight Surgeon, Dr. Jordan, to discuss ways to solve the problem of pilots and aspartame. To date no solutions have been agreed upon.

"4. A warning to PKU children and adults is printed on labels of foods and drinks containing aspartame. However, it must be painfully obvious that these foods named in the petition will *not* be labeled when served in home and other social settings. Those of us who must avoid consumption of aspartame and others who are proven to be a great risk, have no way of knowing and monitoring their consumption of this neurotoxic sweetener. 20 million carriers of the PKU gene are also at risk, and do not know it, in this country.

"5. Many physicians are now warning their pregnant patients to stay away from aspartame during pregnancy. Dr. Louis Elsas of Emory University and Dr. William Partridge of UCLA warn that use of aspartame during pregnancy can cause mental retardation and other birth defects.

"Good Guys" in the Federal Bureaucracy

Not everyone in government is dedicated to defending aspartame products. An occasional sympathetic response from a member of the FDA staff was received by aspartame reactors.

A 64-year-old woman suffered severe headaches, dizziness, insomnia, depression, visual problems, hair changes and ear symptoms while consuming one liter of aspartame-containing soft drinks daily. She wrote:

"It was by the grace of God that I found out what was causing my trouble (aspartame).

"I contacted the FDA and told them what happened to me. I told them health-wise I was in pretty good health before I took aspartame, and if this happened to me it must be happening to others who do not realize what is causing their problem.

"I also asked how they could let this product on the market. His answer to me was, 'This is off the record, but we do not think the public should consume as much as they do.' "

Brave souls in this and other federal agencies could endanger their careers by "going public" with information concerning the toxicity of food additives or toxic wastes. For example, one senior policy adviser in the Occupational Safety Health Administration told Jack Anderson of his anxiety because governmental agencies "act as if chemicals have more rights than people" (*The Palm Beach Post* March 15, 1984, p. A-21).

CHALLENGES TO CONSUMER ADVOCATES

It is unfortunate that one needs to question the motives of "do-gooders" in our highly materialistic society. Even so, the basic motivation of most advocates for aspartame "victims" in the post-thalidomide era seems constructive. They are not—as has been alleged—"health nuts," "frightmongers," "media terrorists," or persons harboring paranoid attitudes against business or science.

In the majority of instances, such intent to protect the public has been altruistic. It is heightened by encounters with bureaucratic and corporate disinterest or arrogance. Consumer advocates decry the proposition that the marketplace should serve as the ultimate correcting force for public ire. Some of their challenges will be reviewed.

"Aspartame Is Everywhere"

Aspartame reactors are virtually unanimous in their concern about the unchecked flooding of their stores and pharmacies with products containing this chemical. As restaurant chefs increasingly replace corn syrup with aspartame products, some hesitate to dine out. (Nathan Pritikin averred that eating in a restaurant is like entering the enemy's camp.) Most reactors are appalled at the foisting of aspartame preparations upon infants and children in the form of "delicious" medicine and vitamins.

Failure of Consumers to Receive Adequate Information from the FDA and Industry

Consumer advocates have attempted to provide information when the public did not receive replies to legitimate queries concerning their reactions to aspartame products. *One out of four* aspartame reactors completing the survey questionnaire had written or called the FDA. Most failed to obtain a satisfactory reply. Moreover, this agency had no "hotline" or mechanism for triaging such calls and letters as of 1988, despite the magnitude of consumer complaints.

The producers of aspartame products usually respond to complaints from consumers. The information they send, however, tends to be biased or too general in nature to benefit the correspondent.

A woman wrote me, "I know beyond a shadow of a doubt that aspartame has bothered me." She had suffered "terrible headaches" which disappeared after stopping aspartame products. Her ensuing experiences in seeking further information was described in this letter.

"I had not read or heard anything adverse about aspartame—this is not why I stopped using it. I just tried to think back when my headaches started and what I was doing differently. The weather had gotten warmer and besides I had started to drink pop. I immediately stopped using any form of aspartame to see if this was the cause.

"It has been over a month now and I have had no headaches at all.

"I called the company who makes aspartame. They referred me to another company to get information about aspartame. I did get a large stack of data, but nothing to really answer my inquiry.

"I called a large producer to inquire as to whether they might bring back saccharin in their drinks—but they wouldn't answer."

Some other examples of responses and dogmatic pronouncements that heightened the resentment of consumers and their advocates are cited.

- I have seen some of the data and statements on the "Nutra-Sweet Consumer Medical Complaint Report" supplied by aspartame reactors who documented in detail their complaints after ingesting aspartame products. The associate medical director replied to one such sufferee

    "Thank you for the information you provided regarding your complaint. I appreciate the time you took to supply me with more details of your medical history and symptoms. I am presently evaluating these data and will incorporate this information into our records."

- Senator Metzenbaum (1988) stated at a Senate hearing:

    "We need new, independent tests of safety. We don't need the company, or non-profit institutes fronting for the company, telling us that this product is safe."

- Dr. Peter Dews (1988), a Ph.D. psychologist and consultant for the International Life Sciences Institute (ILSI), told a Senate hearing that "it is possible to confuse the issue (for the public) by issuing too much information." Yet, he had neither personally researched nor published on aspartame. He also admitted on direct questioning that this Institute is funded by an aspartame manufacturer and the soft drink industry. Gregory Gordon (1988) reported that Dr. Dews had received payments of $31,000 from ILSI during 1984.

Misleading Information and Reassurance

Casual reassurance and misleading information about aspartame-containing products—whether from the FDA, the manufacturer or physicians—pose serious obstacles for consumers and their advocates. This might be regarded as another aspect of the umbrella phenomenon called "canceling reality."

- The Senate hearing held on November 3, 1987 revealed that some callers had been told by the FDA there was "no problem" with aspartame-containing products.
- An FDA consumer safety officer stated, "We concluded that there was no cause-and-effect relationship between the complaints and aspartame . . . There are always individual people who have idiosyncratic reactions to different substances" (*The Miami Herald* July 31, 1986, p. PB-4).
- A patient with fructose intolerance inquired about the safety of aspartame. She was advised, "You shouldn't have a problem with it." She proceeded to use it . . . and suffered a severe reaction.
- A member of Senator Metzenbaum's staff, calling as an individual, received assurance that children could safely consume large amounts of aspartame products (*Congressional Record-Senate* May 7, 1985, p. S5491).
- Physician consultants and scientists allegedly have been hired for large fees by corporate PR firms to defend aspartame in media interviews and other public forums (see Chapter 32). According to Gregory Gordon (1988), others have received "courtesy honorariums" or were wooed with luxurious retreats "to shut up."

Inadequate or Misleading Labels and Ads

The incomplete labeling of aspartame-containing products and "practiced deception" through misleading innuendo have been mentioned earlier. Existing labels also may confuse persons who seek to avoid sugar. For example, many use a popular tabletop aspartame sweetener on the assumption it contains "no sugar"—a reference to sucrose or table sugar. Its label, however, lists "dextrose with dried corn syrup" as the *first* ingredient. (Although it is chemically different from table sugar or sucrose, dextrose is indeed a sugar in all respects.)

Corporate Intimidation

Few citizens, publishers or elected officials are willing to champion the cause of aspartame reactors. This could risk being challenged by a billion-dollar industry with enormous biopolitical clout.

The PR Barrage

Television promos for aspartame-containing beverages and foods may "try men's souls."

Parents and health professionals have reason to worry about the potential hazards of aspartame products when used by pregnant women (Chapter 23), children (Chapter 24), and weight-conscious women (Chapter 16). They are mindful of the lasting impact of "memes" (Chapter 24) initiated by television commercials. This is encapsuled in the quip that children now learn their NBCs and CBSs before their ABCs. Yet, the sophisticated media blitz promoting these products remains overwhelming.

The inference that aspartame is a "natural" or "organic" product justifies consumer outrage. It cannot be effectively addressed, however, until the FDA or some other federal agency rules that such advertising is misleading. A registered nurse with severe reactions to aspartame products wrote:

> "In a state of innocence, we are poisoning our bodies with supposedly 'safe substances' that are advertised as natural. However, in truth, it is produced in a factory and is not extracted from natural substances such as 'milk and bananas' as advertised."

Saccharin—The Scapegoat

Aspartame reactors, even when well informed, remain reluctant to consume saccharin because of the mandated warning on labels: "Use of this product may be hazardous to your health. This product contains saccharin which has been determined to cause cancer in laboratory animals."

In my opinion, the continuation of such labeling represents bureaucratic arrogance. There is still no valid evidence for cancer causation by saccharin. The same applies to cyclamate.

LEGAL PRECEDENTS

Some consumer advocates have attempted to obtain injunctions barring the further distribution of aspartame products on the basis of an imminent public health hazard (Chapter 30). James S. Turner, an attorney representing the Community Nutrition Institute, filed such a petition to the Food and Drug Administration on July 17, 1986 requesting "Reconsideration and/or Repeal" on this basis. His efforts proved futile.

The "Implied Warranty" Issue

A few aspartame reactors entered into litigation relative to an "implied warranty" concerning the safety of aspartame-containing products.

One insulin-dependent diabetic developed seizures after consuming several aspartame-containing products. He believed that a product liability statute had been violated. The manufacturer was sued for breach of an implied warranty that these sweeteners were "fit for ordinary use" (*The Grand Rapids Press* March 6, 1986 p. C–6).

The following considerations are pertinent.

- In arguing an alleged breach of implied warranty in product liability litigation, the plaintiff must prove that (1) a defect existed in the product alleged to have caused injury at the time the supplier parted possession with it, and (2) nothing altered the product following its sale. In the case of aspartame, this would seem to focus any blame on the agency or agencies originally licensing it.
- Article 2 of the Uniform Commercial Code addresses the inference of warranty by advertisements. It asserts that affirmations of fact and promises could create a warranty when the seller makes statements about goods to potential buyers. Such warranties concerning safety apply to a wide variety of products that humans ingest, or with which they come in contact.
- It has been asserted: "Mere compliance with federal or state

regulatory labeling requirements does not preclude a jury from finding that additional warnings should have been given by manufacturer" (*Ferebee v. Chevron Chemical Company* 1984).

Pertinent Legal and Judicial Issues

The complexity of litigation involving allegations of long-term injuries from exposure to a toxic product is convincingly illustrated by the magnitude of law suits concerning Agent Orange. A recent ruling has relevance to the present subject.

A United States District Court in California ruled in *Nehmer v. United States Veterans Administration* (1989) that the defendants had failed to comply with the Veterans' Dioxin and Radiation Exposure Compensation Standards Act (Dioxin Act) by erroneously interpreting two provisions of the Act in its summary judgment, and remanded the case to the VA for further proceedings. This court noted that the VA had wrongfully required that the scientific evidence demonstrated a "cause and effect" relationship between Agent Orange exposure and claimed diseases, instead of using the less demanding standard that there be a "statistical association" between Agent Orange exposure and claimed diseases. Moreover, several declarations of plaintiff expert witnesses stated that the exclusion of the animal studies had resulted in an incomplete review of the scientific literature.

In litigation involving the liability of manufacturers for products, proof of causation is a prerequisite for recovery under any legal theory of liability. An acceptable basis for causation involves not only epidemiology, but also the so-called "signature effect" or syndrome as accurately documented by an expert witness.

The enormous escalation of litigation involving liability of manufacturers and sellers of products that allegedly caused harm, especially with an actual or implied warranty of safety, has resulted in an enormous expansion of tort law and cases—especially during the past decade.

The issue of the "defective" nature of a product—whether a chemical or a machine—has been probed by many courts for purposes of imposing strict liability. Three functionally distinct types of defect exist—namely, those involving defective design, defects in construction, and inadequate warnings. In *Phillips v. Kimwood Machine Co.* (1974), the Supreme Court of Oregon noted

> "a dangerously defective article would be one which a reasonable person would not put into the stream of commerce *if he had knowledge of its harmful character.* The test, therefore, is whether the seller would be negligent if he sold the article *knowing of the risk involved.* Strict liability imposes what amounts to constructive knowledge of the condition of the product."

It has been ruled that a corporate defendant can be held liable if he is negligent in testing his product and fails to discover or to reveal that it is unsafe. A case in point is *Schenebeck v. Sterling Drug, Inc.* (1970).

The Duty To Inform

There are multiple laws and rulings pertaining to the duty of manufacturers and producers to warn of all risks, either known or reasonably foreseeable. A few are cited as examples.

- The continuing duty of the manufacturer to keep abreast of scientific developments—as in *McEwen v. Ortho Pharmaceutical Corp.* (1974).
- The providing of warning with sufficient intensity to invite safety commensurate with potential risks by a reasonable person—as in *Tampa Drug Co. v. Waite* (1958).
- The duty to warn of newly-discovered risks in the form of labeling changes and various timely methods for communicating such information to the medical community—as in *Sterling Drug v. Yarow* (1969).

The nature of the duty to warn with an appropriate degree of intensity for a reasonable person has been emphasized by courts.

- In *Spruill v. Boyle-Midway, Inc.* (1962), it was asserted:

> *To be of such character the warning must embody two characteristics: first, it must be in such form that it could reasonably be expected to catch the attention of the reasonably prudent man in the circumstances*

of its use; secondly, the content of the warning must be of such a nature as to be comprehensible to the average user and to convey a fair indication of the nature and extent of the danger to the mind of a reasonably prudent person. (Italics added)

- In *D'Arienzo v. Clairol, Inc.* (1973); the Supreme Court of New Jersey stated: "Care is to be taken that the reasonable man be not endowed with attributes which properly belong to a person of exceptional perspicuity and foresight."
- In *Barson v. E. R. Squibb & Sons, Inc.* (1984), the Supreme Court of Utah emphasized:

. . . the drug manufacture is held to be an expert in its particular field and is under a "continuous duty . . . to keep abreast of scientific developments touching upon the manufacturer's product and to notify the medical profession of any additional side effects discovered from its use." The drug manufacturer is responsible therefore for not only "actual knowledge gained from research and adverse reactions reports," but also for "constructive knowledge as measured by scientific literature and other available means of communication."

The Vulnerability of Governmental Agencies

Regulatory agencies that license potentially toxic products, even with knowledge of such purported toxicity, heretofore have been relatively immune against liability by consumers and consumer advocates under a provision in the Federal Tort Claims Act (FTCA).

This situation changed when the U.S. Supreme Court ruled on June 13, 1988—in the case of *Berkovitz v. United States*—that the government *can* be sued for injuries caused by government-regulated products if the regulators who approved them negligently violated government policies. (The case focused on approving release of a particular lot of oral poliomyelitis vaccine to the public that allegedly caused a severe case of polio.) In a 9–0 decision, the Court rejected the argument that government is immune from liability for "any and all acts arising out of the regulatory programs of federal agencies."

In view of the importance of this decision for consumers and consumer advocates, some background information is offered.

- The FTCA generally authorizes suits against the United States for damages

  "for injury or loss of property, or personal injury or death caused by the negligent or wrongful act or omission of any employee of the Government while acting within the scope of his office or employment, under circumstances where the United States, if a private person, would be liable to the claimant in accordance with the law of the place where the act or omission occurred" (28 U.S.C. 1346[b]).

- The FTCA included a number of exceptions to this broad waiver of sovereign immunity, the one relevant to this case providing that no liability shall lie for

  "(any) claim . . . based upon the exercise or performance or the failure to exercise or perform a discretionary function or duty on the part of the federal agency or an employee of the Government, whether or not the discretion involved be abused" (28 U.S.C. 2680[a]).

  This provision reflected the line of demarcation between the willingness of Congress to impose tort liability upon the United States on the one hand, and its desire to protect certain governmental activities from exposure to suit by private persons (*United States v. Varig Airlines* 467 U.S.).

- The Division of Biologic Standards (DBS) is required to comply with certain statutory and regulatory provisions before issuing a license. The Public Health Service Act asserts:

  "Licenses for the maintenance of establishments for the propagation or manufacture and preparation of products (including polio vaccines) may be issued only upon a showing that the establishment and the products for which a license is desired meet standards, designed to insure the continued safety, purity, and potency of such products, prescribed in regulations, and licenses for new products may be issued only upon a showing that they meet such standards."

- Furthermore, this Act required the agency to receive all the mandated data from the manufacturer, and to determine the safety of the product before issuing a license.

An appellate court may grant summary judgment in part to aplaintiff, and remand the case for a new trial for further proceedings by a regulatory agency of the federal government. In *Nehmer v. United States Veterans Association* (1989), a United States District Court found that the defendants had failed to comply with the Dioxin and Radiation Exposure Compensation Standards Act by erroneously interpreting two provisions of the Act.

CHAPTER THIRTY-FIVE

Alternatives to Aspartame

*Almost every second store and shop in our villages
and cities is a candy store, and common sense and
common observation knows well enough the
morbid results . . . How many patients have
blessed me for the suggestion (to stop all sweets),
and have traced to the continued rules their
reinstated health and enjoyment of life.*
        — DR. GEORGE M. GOULD
                            (1910)

MANY PATIENTS AND APPARENT reactors to aspartame-containing products have asked me about ways of satisfying their "sweet tooth" without using aspartame products. This is a uniquely important decision for persons with diabetes mellitus and severe hypoglycemia who ought to avoid or minimize table sugar (sucrose), as well as aspartame. Sir William Osler, the great physician, properly warned at the turn of the century: "Cookery simulates the disguise of medicine and pretends to know what food is best for the body."

This matter is complicated by the fact that considerable controversy exists as to what actually constitutes an optimum diet for such patients. For example, a "high protein diet" is generally recommended for hypoglycemic individuals. But excess protein can release considerable insulin—thereby aggravating or provoking low blood sugar attacks.

Because of a decline in sugar consumption, sugar cane growers, refiners and processors throughout the country have launched a multimillion-dollar campaign to focus on the "sweet, pure and natural" aspects of sugar (*News/Sun-Sentinel* of Fort Lauderdale, February 21, 1988, D–1). In my opinion, the success of such promotion could exacerbate the problems of diabetes, hypoglycemia and obesity for reasons discussed earlier.

The advice I generally offer is: "Be cautious. Read labels. Think before sweetening. And don't be fooled by appealing commercials, the fine print, or slogans on labels that read 'diet,' 'lite,' 'low-calorie,' or 'sugar-free,' and presumed sweetener breakthroughs that set you up as a possible guinea pig." The last comment should come as not surprise to the reader of this book. Food technologists are currently working at a feverish pace to produce other synthetic sweeteners.

GENERAL OPTIONS

The following generalizations reflect my experience. They are *not* intended to replace the recommendations of physicians treating individual patients.

Water

Some aspartame sufferees finally resign themselves to choosing "the lesser of several evils:" water.

> A 41-year-old woman with multiple and severe reactions to aspartame products stated, "Many times I long for a soft drink— especially on a hot day or when we go out for pizza. Now many companies use aspartame, and I can't have any. I drink only water."

Whether from the tap or bottled as "pure" (possibly misleading), water is better than any "diet" drink. If desired, lemon, lime or various sugar-free flavorings may be added.

Seltzer water should be used with discretion by persons having a hiatal hernia or esophageal reflux.

Unsweetened Beverages

Most persons can consume modest amounts of tea, coffee (regular or decaffeinated), unsweetened tea mixes, herb teas, and other beverages. Lemon, lime, various flavorings, saccharin and cyclamate may be added as desired.

A few comments about *caffeine* are pertinent.

Patients with hypertension, heart problems, diabetes and hypoglycemia should be prudent about their intake of caffeine. This includes saccharin-sweetened colas.

Caffeine enhances the taste of food, especially among older persons whose sense of taste tends to become dulled. Problems could arise from certain sweeteners when such enjoyment results in excessive consumption—and a possible vicious cycle of side effects.

Saccharin, Cyclamate and Acesulfame

In my opinion, a reasonable intake of saccharin and cyclamate—whether as tablets, drops, or presweetened soft drinks—is generally safe (Chapter 32). Some prefer the calcium form of saccharin (if obtainable) owing to the absence of an aftertaste.

I would offer another important caveat to aspartame sufferees: *constantly* remain alert to the possibility that producers may substitute aspartame for saccharin to meet their competition. Under these circumstances, they have been known to use up supplies of old labels that do not indicate the presence of aspartame.

The FDA is currently considering the reinstatement of *cyclamate* after banning it during 1969. In the intervening two decades, there has been no evidence for cancer or other disease linked to its use in countries where it continues to be sold. Although the FDA failed to demonstrate that cyclamate causes cancer, there is still concern by some as whether cyclamate might be a co-carcinogen because of its extensive metabolism and byproducts.

Acesulfame potassium (or acesulfame K) was approved for use by the FDA as a safe artificial sweetener in July 1988. It is a derivative of acetoacetic acid. Unfortunately, several potential problems associated with the use of acesulfame have been raised. They are based largely on animal studies since testing on humans remains limited.

- It stimulates insulin secretion in a dose-dependent fashion, thereby possibly aggravating reactive hypoglycemia ("low blood sugar attacks").
- It apparently produced lung tumors, breast tumors, rare types

of tumors in other organs (such as the thymus gland), several forms of leukemia, and chronic respiratory disease in several rodent studies, even when less-than-maximum doses were given. Accordingly, the Center for Science in the Public Interest formally petitioned on August 29, 1988 for a stay of its approval by the FDA on August 29, 1988 because of "significant doubt" about the safety of acesulfame potassium.

Fruit Juices

"Unsweetened" fruit juices usually can serve as replacements for aspartame beverages. But some may create problems for diabetic and hypoglycemic patients. For example, a glass of orange juice tends to release more insulin than eating an orange. Similarly, "unsweetened" prune juice or grape juice may contain considerable "natural" sugar.

Fruits

Pears, peaches, strawberries and grapefruit usually serve as satisfactory snacks. On the other hand, fruits containing considerable carbohydrate (e.g., mangos, apricots, prunes, raisins) should be avoided by hypoglycemic patients . . . or taken with adequate fat and protein.

Laxatives

Psyllium hydrophilic mucilloid preparations are commonly used for managing constipation and the "irritable bowel syndrome." Since several popular brands contain either dextrose or aspartame, they should be avoided by diabetics, patients prone to hypoglycemia, and aspartame reactors. Some (such as Hydrocil Instant® by Rowell) contain neither dextrose or aspartame.

Sugar as an Option

I have long recommended that patients with diabetes and reactive hypoglycemia avoid or minimize sugar (sucrose), especially when unsuspected in many "foods of commerce." Other undesirable aspects of sugar consumption have been reviewed in my previous publications (Roberts 1967a, 1968b, 1969c). They include its contributory role to excessive insulin release (especially in obesity), and elevation of the blood triglyceride (fat) and cholesterol concentrations.

Accordingly, the proper answer to a question often asked me by apparent aspartame reactors, "Doctor, which is worse for me . . . sugar or aspartame?", generates anguish. My present opinion is that sugar, taken in minimum amounts, is preferable. Most of my diabetic and hypoglycemic patients seem to fare relatively well so long as they also ingest adequate protein and fat, and take small mid-afternoon and bedtime snacks. Moreover, many find that their sweet tooth tends to have been blunted following prior abstention from sugar.

The First—and Last—Word

AT MY FIRST PRESS conference on reactions to aspartame products, I suggested that the main benefit of bringing the aspartame story to public attention was that of permitting other physicians, investigators and even consumers to verify or challenge my observations. I had no aspirations of becoming a majority-of-one physician counterpart of Rambo for "aspartame victims."

I have made a desperate attempt to maintain my professional objectivity and integrity, without intent of malice, in pursuing this clinical and scientific probe.

Years later, the following statement made at the initial conference still seems appropriate.

> "I think it would be a tragedy if this issue is ignored since we could be inviting disastrous medical, psychological and neurological problems. I hope I'm wrong. But let's look at the problem NOW instead of in five or 10 years when we might be having a medical plague on our hands."

REFERENCES

Abrams, F. R. 1986. Patient advocate or secret agent? *Journal of the American Medical Association* 256:1784–85.

Adamkiewicz, V. W. 1963. Glycemia and immune responses. *Canadian Medical Association Journal* 88:806.

Adamsons, K. 1973. Obesity and Fad Diets. Testimony for the U.S. Senate Select Committee on Nutrition and Human Needs, April 12.

Alexander, E. 1986. Cited by *Medical Tribune* November, p.ll.

Altman, L. 1988. Leaked story raises questions about medical journal's power. *The Palm Beach Post* January 29, p.E–5.

American Cancer Society 1986. Cancer Prevention Study II: An epidemiological study of lifestyles and environment. *CPS II Newsletter*, Spring, 4/1:3.

American Diabetes Association 1987. Position Statement: Use of noncaloric sweeteners. *Diabetes Care* 10:526.

Anderson, J. W., N. Zettwoch, and T. Feldman, et al. 1988. Cholesterol-lowering effects of psyllium hydrophilic mucilloid for hypercholesterolemic men. *Archives of Internal Medicine* 148:292296.

Antener, L., A. M. Verwilghen, C. Vangeert, and J. Mauron 1981. Biochemical study of malnutrition. 5. Metabolism of phenylalanine and tyrosine. *International Journal of Vitamin and Nutrition Research* 51:296–306.

Atkinson, R. L. 1987. Opioid regulation of food intake and body weight in humans. *Federation Proceedings* 46(1):178–182.

Bailar, J. C. III, A. M. Finkel, E. K. Silbergeld, and D. Hattis 1989. Cancer-causing substances in food, drugs, and cometics. (Letter) *New England Journal of Medicine* 320:935.

Barson v. E. R. Squibb & Sons, Inc., 682 P.2d 832 (Utah 1984).

Baumbach, F. L., P. A. Cancilla, G. Martin-Amat, T. R. Tephly, K. E. McMartin, A. B. Maker, and M. S. Hayreh 1977. Methyl alcohol poisoning. IV. Alterations of the morphological findings of the retina and the optic nerve. *Archives of Ophthalmology* 95:1859–1865.

Beckman, H. B. and R. M. Frankel 1984. The effect of physician behavior on the collection of data. *Annals of Internal Medicine* 101:692–696.

Bender, D. A. 1985. *Amino Acid Metabolism*. Chichester/New York, Wiley, 2nd Edition.

Bennett, I. L., Jr., T. C. Nation, and J. F. Olley 1952. Pancreatitis in methyl alcohol poisoning. *Journal of Laboratory and Clinical Medicine* 40:405–409.

Bennett, I. L. Jr., F. H. Carey, G. L. Mitchell, Jr., and M. N. Cooper 1953. Acute methyl alcohol poisoning. A review based on experiences in an outbreak of 323 cases. *Medicine* 32:431.

Berkovitz v. United States, 100 L.Ed.2d 531 (1988).

Bernard, C. 1927. *Introduction to the Study of Experimental Medicine*. MacMillan Publishing Company, New York, p.82

Bigby, M., S. Gick, H. Gick, and K. Arndt 1986. Drug-induced cutaneous reactions. *Journal of the American Medical Association* 256:3358–3360.

Blundell, J. E. and A. J. Hill 1986. Paradoxical effects of an intense sweetener (aspartame) on appetite. *The Lancet* 1:1092–1093.

Blundell, J. E., A. J. Hill, and P. J Rogers 1987. Effects of aspartame on appetite and food intake. In *Proceedings of the First International Meeting on Dietary Phenylalanine and Brain Function*, edited by R. J. Wurtman and E. Ritter-Walker, Washington, D.C., May 8–10, pp.299–309.

Boehm, M. F. and J. L. Bada 1984. Racemization of aspartic acid and phenylalanine in the sweetener aspartame at 100°C. *Proceedings of the National Academy of Sciences USA* 81: 5263–5266.

Bombeck, E. 1988. Give me some real motivation to be thin. *The Palm Beach Post* December 11, p.F–12.

Bortz, W. P. M., P. Howat, and W. L. Holmes 1969. The effect of feeding frequency on diurnal plasma free fatty acids and glucose levels. *Metabolism* 18:120.

Brayne, C. and D. Calloway 1988. Normal ageing, impaired cognitive function, and senile dementia of the Alzheimer's type: a continuum? *Lancet* 1:1265–1266.

Brown, E. S., H. A. Waisman, M. A. Swanson, R. E. Colwell, M. E. Banks, and T. Gerritson 1973. Effects of oral contraceptives and obesity on carrier tests for phenylketonuria. *Clinica Chimica Acta* 44:183–192.

Brown v. Superior Court, 751 P.2d 470 (Cal. 1988).

Brown, W. D. 1967. Present knowledge of protein nutrition, Part 3, *Postgraduate Medicine* April, pp.119–126.

Bryan, G. T. 1894. Artificial sweeteners and bladder cancer: Assessment of carcinogenicity of aspartame and its diketopiperazine derivative in mice. In *Aspartame: Physiology and Biochemistry*, edited by L. D. Stegink and L. J. Filer, Jr., Marcel Dekker, Inc., New York, pp.321–348.

Burget, S. L., D. W. Andersen, and L. D. Stegink, et al. 1984. Aspartame and phenylalanine methyl ester metabolism by the porcine gut. (Abstract) *Clinical Research* 32:762.

Bryson, D. D. 1981. Health hazards of formaldehyde. *The Lancet* 1:1263.

Caballero, D. and R. J. Wurtman 1987. Control of plasma phenylalanine levels. In *Proceedings of the First International Meeting on Dietary phenylalanine and Brain Function,* edited by R. J. Wurtman and E. Ritter-Walker, Washington, D.C. May 8–10, pp.9–23.

Caldecott, R. 1961. Cited by *Science Newsletter* November 18.

Callaway, C. W. 1986. Nutrition. *Journal of the American Medical Association* 256:2097–2098.

Carlson, H. E. 1989. Prolactin stimulation by protein is medicated by amino acids in humans. *Journal of Clinical Endocrinology and Metabolism.* 69:7–14.

Ceci, F. et al. 1986. Oral 5-hydroytryptophan promotes weight loss in bulimic obese subjects. *Clinical Research* 34:866A.

Chapman, A. G., B. Engelsen, and B. S. Meldrum 1987. 2-Amino-7- phosphonoheptanoic acid inhibits insulin-induced convulsions and striatal aspartate accumulation in rats with frontal cortical ablation. *Journal of Neurochemistry* 49:121–127.

Choi, T. B., J. Yang, and W. M. Pardridge 1986. Phenylalanine transport at the human-blood-brain barrier: Studies with isolated brain capillaries. *Diabetes* 35 (Suppl.l): 196.

Christensen, H. N. 1987. Dual role of transport competition in amino acid deprivation of the nervous system. In *Proceedings of the First International Meeting on Dietary Phenylalanine and Brain Function,* edited by R. J. Wurtman and E. Ritter-Walker, Washington, D.C., May 8–10, pp.95–104.

Clark, L. H., Jr. 1987. How protectionism soured the sugar market. *The Wall Street Journal* November 4, p.36.

Cloninger, M. R. and R. E. Baldwin 1970. Aspartylphenylalanine methyl ester: A low calorie sweetener. *Science* 170:81.

Collings, M. 1988. Testimony for the Committee on Labor and Human Resources, U.S. Senate, Hearing on *"NutraSweet"—Health and Safety Concerns,* November 3, 1987. 83–178, U.S. Government Printing Office, Washington, pp.305–307.

Congressional Record-Senate 1985. Saccharin Study and Labeling Act Amendments of 1985. May 7, pp.S5489–5516.

Congressional Record-Senate 1985. Aspartame Safety Act of 1985. August 1, pp.S10820–10847.

Coon, J. J. 1970. Food toxicology: Safety of food additives. *Modern Medicine* November 30, p.105.

Cornell, R. G., R. A. Wolfe, and P. G. Sanders 1984. Aspartame and brain tumors: Statistical issues. In *Aspartame: Physiology and Biochemistry,* edited by L. D. Stegink and L. J. Filer, Jr., Marcel Dekker, Inc., New York, pp.459–479.

Council of Scientific Affairs: Aspartame: Review of safety issues. 1985. *Journal of the American Medical Association* 254:400–402.

Coyle, J. 1988. More dimensions for glutamate toxicity. Cited by *Science* 242:1509–1510.

Craft, I. L. and T. J. Peters 1971. Quantitative changes on plasma amino acids induced by oral contraceptives. *Clinical Science* 41:301–307.

Cross, A. J., D. Slater, and M. Simpson, et al. 1987. Sodium dependent D-(3H) aspartate binding in cerebral cortex in patients with Alzheimer's and Parkinson's diseases. *Neurosciences Letter* 79:213–217.

Curran, D. A. and J. W. Lance 1964. Clinical trial of methysergide and other preparations in management of migraine. *Journal of Neurology, Neurosurgery & Psychiatry* 27:463.

D'Adamo, A. F., Jr. and F. M. Yatsu 1966. Acetate metabolism in the nervous system. N-acetyl-L-aspartic acid and the biosynthesis of brain lipids. *Journal of Neurochemistry* 13:961–965.

D'Arienzo v. Clairol, Inc., 310 A2d 106. (1973).

Dews, P. 1988. Testimony for the Committee on Labor and Human Resources, U.S. Senate, Hearing on *"NutraSweet"—Health and Safety Concerns* November 3, 1987. 83–178, U.S. Government Printing Office, Washington, pp.374–376.

Dhont, J. L., G. Kapatos, M. Parniak, H. Wilgus, and S. Kaufman 1982. Biopterin metabolism and phenylalanine hydroxylase activity during early liver regeneration. *Biochemical and Biophysical Research Communications* 106:786–793.

Dolan, G. and C. Godin 1966. Phenylalanine toxicity in rats. *Canadian Journal of Biochemistry* 44:143–145.

Dow-Edwards, D. L. and G. Deibler 1989. Developmental toxicity of aspartame: Effects on plasma amino acid levels. Abstract 323. *Transactions of the American Society for Neurochemistry*.

Drake, M. E. 1986. Panic attacks and excessive aspartame ingestion. *The Lancet* 2:631.

Dunbar, P. V. 1950. Statement to the U.S. House of Representatives Select Committee to Investigate the Use of Chemicals in Food Products, 81st Congress, 2nd Session.

Ehrlich, J. P. 1987. Cholesterol awareness. (Editorial) *Medical Tribune* November 11, p.32.

Elsas, L. J., II and J. F. Trotter 1987. Changes in physiological concentrations of blood phenylalanine-produced changes in sensitive parameters of human brain function. In *Proceedings of the First International Meeting on Dietary Phenylalanine and Brain Function*, edited by R. J. Wurtman and E. Ritter-Walker, Washington, D.C., May 8–10, pp.263–273.

Erlanson, P., et al. 1965. Severe methanol intoxication. *Acta Medica Scandinavica* 177:393.

Farkas, C. S. and C. E. Forbes 1965. Do non-caloric sweeteners aid patients with diabetes to adhere to their diets? *Journal of the American Dietetic Association.* 46:482–484.

Felbeck, H. and S. Wiley 1987. Free D-amino acids in the tissues of marine bivalves. Cited by Man and Bada (1987).

Felig, P., E. Marliss, and G. F. Cahill 1969. Plasma amino acid levels and insulin secretion in obesity. *New England Journal of Medicine* 281:811–816.

Ferebee v. Chevron Chemical Company, 736 F.2d 1529 (D.C. Cir.), *Cert. denied,* 469 U. S. 1062 (1984).

Ferguson, J. N. 1985. Interaction of aspartame and carbohydrates in an eating-disordered patient. *American Journal of Psychiatry* 142:271.

Fernstrom, J. D. 1987. Letter. *American Journal of Nutrition.* 45:801–803.

Finkelstein, M. W., T. Daabees, L. D. Stegink, and A. E. Applebaum 1983. Correlation of aspartame dose, plasma dicarboxylic amino acid concentration, and neuronal necrosis in infant mice. *Toxicology* 29, 109–119.

Finley, J. W. and D. E. Schwass 1983. *Xenobiotics in Foods and Feeds. American Chemical Society Symposium Series,* No. 234, Washington, D.C.

Floyd, J. C., Jr., S. Fajans, J. W. Conn, R. F. Knopf, and J. Rull 1966. Stimulation of insulin secretion by amino acids. *Journal of Clinical Investigation* 45:1487–1502.

Floyd, J. C., S. S. Fajans, and S. Pek, et al 1970. Synergistic effect of certain amino acid pairs upon insulin secretion in man. *Diabetes* 19:102–108.

Forbes, A. L. 1986. Dimensions of the issue of explicit health claims on food labels. *American Journal of Clinical Nutrition* 43:629–635.

Forney, R. B. and F. W. Hughes 1968. *Combined Effects of Alcohol and Other Drugs.* Charles C Thomas, Springfield, pp.92–95.

Francot, P. and P. Geoffroy 1956. Le methanol dans les jus de fruits, les boissons, fermentees, les alcools et spiritueux. *Rev. Ferment, Inc. Ailment.* 11:279–287.

Freedman, D. X. and L. D. Grouse 1986. Physicians and the mental illnesses: The nudge from ADAMHA. *Journal of the American Medical Association* 255:2485–2486.

Fürst, P., A. Alvesstrand, and J. Bergström 1980. Effects of nutrition and catabolic stress on intracellular amino acid pools in uremia. *American Journal of Clinical Nutrition* 33:1387–1395.

Genazzani, A. R. and F. Facchinette, et al. 1986. Hyperendorphinemia in obese children and adolescents. *Journal of Clinical Endocinology & Metabolism* 62:36–40.

Gilger, A. P., I. S. Farkas, and A. M. Potts 1959. Studies on the visual toxicity of methanol X. Further observations on the ethanol therapy of acute ethanol poisonings in monkeys. *American Journal of Ophthalmology* 48:153–161.

Go, V. L. M., A. F. Hoffman, and W. H. J. Summerskill 1970. Pancreozymin bioassay in man based on pancreatic enzyme secretion: Potency of specific amino acids and other digestive products. *Journal of Clinical Investigation* 49:1558–1564.

Goldstein, D. S., R. Udelsman, and G. Eisenhofer, et al. 1987. Neuronal source of plasma dihydroxyphenylalanine. *Journal of Clinical Endocrinology and Metabolism* 64:856–861.

Goodman, L. S. and A. Gilman 1980. *The Pharmacological Basis of Therapeutics* 6th Ed. MacMillan Publishing Company, New York, pp.386–387.

Gordon, G. 1988. UPI investigative report: NutraSweet: Questions swirl. Committee on Labor and Human Resources, U.S. Senate, Hearing on *"NutraSweet"—Health and Safety Concerns* 83–178, U.S. Government Printing Office, Washington, pp.483–510.

Gordon, G. S. and H. W. Elliott 1947. Action of diethylstilbestrol and some steroids on respiration of rat brain homogenates. *Endocrinology* 41:517.

Gould, G. M. 1910. Epilepsy and other diseases due to albumin- starvation and sugar poisoning. *Medical Review of Reviews* July, pp.1–4.

Graves, F. 1984. How safe is your diet soft drink? *Common Cause* Magazine. July/August, pp.24–43.

Grater, W. C. 1969. Cited by *Medical Tribune* August 18, p.lO.

Hagenfeldt, L., I. Bollgren, and N. Venizelos 1987. N-acetylaspartic aciduria due to aspartoacylase deficiency—a new aetiology of childhood leukodystrophy. *Journal of Inherited and Metabolic Diseases* 10:135–141.

Harper, A. E., N. J. Benevenga, and R. M. Wohlhueter 1970. Effects of ingestion of disproportionate amounts of amino acids. *Physiological Reviews* 50:439–448.

Harper, A. E. 1988. Killer french fries: the misguided drive to improve the American diet. *The Sciences* January/February, pp.2128.

Harris, C. 1987. Chelation: Anecdote vs science. *Medical Tribune* September 16, p.14.

Hart, P. 1968. Remarks to National Council of Salesmen's Organization, Inc., New York City, December 2.

Hayreh, M. S., S. S. Hayreh, G. L. Baumbach, P. Cancilla, G. Martin-Amat, T. R. Tephly, K. E. McMartin, and A. B. Makar 1977. Methyl alcohol poisoning III. Ocular toxicity. *Archives of Ophthalmology* 95:1851–1858.

Health and Public Policy Committee 1986. Eating disorders: Anorexia nervosa and bulimia. *Annals of Internal Medicine* 105:790–794.

Heberer, M., H. Talke, K. P. Maier, and W. Gerok 1980. Metabolism of phenylalanine in liver diseases. *Klinische Wochenschrift* 58:11891196.

Hill, A. B. 1965. The environment and disease: Association or causation? *Proceedings of Royal Society of medicine* 589:295–300.

Hodge, H. C. 1963. Research needs in the toxicology of food additives. *Food and Cosmetics Toxicology* 1: September, p.31.

Hoffman, M. 1936. *Heads and Tales.* New York, Charles Scribner's Sons, p.41.

Holden, C. 1986. Depression research advances, treatment lags. *Science* 233:723–726.

Holliday, M. A. 1986. Book review. *New England Journal of Medicine* 315:654.

Homburger, F. 1977. Saccharin and cancer. *New England Journal of Medicine* 297:560–561.

Hommes, F. A. and K. Matsuo 1987. Effect of phenylalanine on brain maturation: implications for the treatment of patients with PKU. In *Proceedings of the First International Meeting on Dietary Phenylalanine and Brain Function,* edited by R. J. Wurtman and E. Ritter-Walker, Washington, D.C., May 8–10, pp.229–236.

Horwitz, D. L., Bauer-Nehrling, and J. K. Cohen 1983. Can aspartame meet our expectations? *Journal of the American Dietetic Association* 83:142146.

Hotchkiss, W. S. 1987. Doctor as patient advocate. *Journal of the American Medical Association* 258:947–948.

Hussein, N. M., R. P. D'Amelia, A. L. Manz, H. Jacin, and W. T. C. Chen 1984. Determination of reactivity of aspartame with flavor aldehydes by gas chromatography, HPLC and GPC. *Journal of Food Science* 49:520–524.

Hutt, P. B. 1973. Safety Regulation in the Real World. Address to the First Academy Forum on the Design of Policy on Drugs and Food Additives, National Academy of Sciences, Washington, D.C., May 15.

Ivanitskiy, A. M. 1973. Evaluation of the content of methanol formed in certain beverages by fermentation hydrolysis of pectin. (Translated from the Russian) Nutrition Institute, USSR Academy of Medical Sciences, Moscow, March 3.

Jaffe, G. J. and T. C. Burton 1988. Progession of nonproliferative diabetic retinopathy following cataract extractions. *Archives of Ophthalmology.* 106:745–749.

Jagenburg, R., R. Olsson, C. G. Regardh, and S. Rödjer 1977. Kinetics of intravenously administered L-phenylalanine with cirrhosis of the liver. *Clinica Chimica Acta* 78:453–463.

Javier, Z., H. Gershberg, and M. Hulse 1968. Ovulatory suppressants, estrogens, and carbohydrate metabolism. *Metabolism* 17:443.

Joachim, J., G. Kalantzis, and H. Delonca, et al. 1987. The influence of particle size on the strain applied on the punches during compression concerning effervescent tablets of aspartame (translation of the French title). *Journal de Pharmacie de Belgique* 42:17–28.

Jobe, P. C. and J. W. Dailey 1987. Role of monoamines in seizure predisposition in the genetically epilepsy-prone rats. In *Proceedings of the First Inter-*

national Meeting on Dietary Phenylalanine and Brain Function, edited by R.J. Wurtman and E. RitterWalker, Washington, D.C., May 8–10, pp.143–160.

Johns, D. R. 1986. Migraine provoked by aspartame. *New England Journal of Medicine* 315:456.

Johnson, P. E. 1970. Cited by *Chemical and Engineering News* March 9.

Johnson, R. 1988. Nutritionists detect a dark side in new world of food substitutes. *The Wall Street Journal* February 3, p.25.

Jones, M. R., J. D. Kopple, and M. E. Swenseid 1978. Phenylalanine metabolism in normal and uremic man. *Kidney International* 14:169177.

Jorgensen v. Meade Johnson Laboratories, Inc. 483 F.2d 237 (1973).

Kahn, E. 1987. Ephraim Kahn replies. *The Journal of Pesticide Reform* 7:37.

Kako, K. 1965. Relationship between endogenous fuel and performance of isolated hearts of fed, starved and alloxan diabetic rats. (Abstract) *Circulation* 31 (II):121.

Kamm, T. 1988. Soft drinks get the hard sell in Europe. *The Wall Street Journal* November 21, p.B-7.

Kassirer, J. P. 1989. Our stubborn quest for diagnostic certainty: A cause of excessive testing. *New England Journal of Medicine* 320:1489–1491.

Kim, K. C., M. D. Tasch, and S. H. Kim 1987. The effect of aspartame on 50% convulsion doses of lidocaine. In *Proceedings of the First International Meeting on Ddietary Phenylalanine and Brain Function*, edited by R. J. Wurtman and E. Ritter-Walker, Washington, D.C., May 8–10, pp.431–435.

Klahr, S. and M. L. Purkerson 1988. Effects of dietary protein on renal function and on the progression of renal disease. *American Journal of Clinical Nutrition* 47:146–152.

Klavins, J. V. 1967. Pathology of amino acid excess: VII. Phenylalanine and tyrosine. *Archives of Pathology* 84:238–250.

Koch, R. and M. Blaskovics 1982. Four cases of hyperphenylalaninemia: Studies during pregnancy and of the offspring produced. *Journal of Inherited and Metabolic Diseases* 5:11–15.

Koehler, S. M., A. Glaros, and E. B. Fennell, et al. 1987. The effects of aspartame consumption on migraine headache. In *Proceedings of the International Meeting on Dietary Phenylalanine and Brain Function*, edited by R. J. Wurtman and E. Ritter- Walker, Washington, D.C., May 8–10, pp.441–447.

Koivusalo, M. 1958. Effect of disulfiram (tetraethylthiuram disulphide) on the elimination rate of methanol. (Abstract). *Quarterly Journal of Studies on Alcoholism* 19:363 (June).

Korte, D. 1989. Is the FDA only guessing? *The Wall Street Journal* January 26, p.A-15.

Krebs, H. H. 1935. Metabolism of amino acids. III. Deamination of amino acids. *Biochemical Journal* 29:1620–1644.

Krebs, H. H. 1948. The D- and L-amino acid oxidases. *Biochemical Society Symposium* 1:2–19.

Kreusi, M. J. P., J. L. Rapoport, and E. M. Cummings, et al. 1986. Sugar or aspartame: Effects on aggression and activity. Presented at 139th Annual Meeting of the American Psychiatric Association, Washington, D.C., May 12.

Kulczycki, A., Jr. 1986. Aspartame-induced urticaria. *Annals of Internal Medicine* 104:207–208.

Kulczycki, A., Jr. 1987. Aspartame allergy. *Allergy Observer* June, p.6.

Landau, R. L. and K. Lugibihl, 1967. The effect of progesterone on the concentration of plasma amino acids in man. *Metabolism* 16:1114–1122.

Lehmann, W. D. and H. C. Heinrich 1986. Impaired phenylalanine-tyrosine conversion in patients with iron-deficiency anemia studied by a L-(^{2}Hs) phenylalanine-loading test. *American Journal of Clinical Nutrition* 44:468–474.

Leibel, R. L. and J. Hirsch 1987. Site-and sex-related differences in adrenoreceptor status of human adipose tissue. *Journal of Clinical Endocrinology & Metabolism* 64:1205–1210.

Le Moan, M. G. 1956. Methanol in fruit juices: Methanol analysis in officinal fruit juices and in various preparations based on vegetable juice. (Translation) *Annales Pharmaceutiques Francaises* 14:470–475.

Levy, H. L. and S. E. Waisbren 1983. Effects of untreated maternal phenylketonuria and hyperphenylalananemia on the fetus. *New England Journal of Medicine* 309:1269–1274.

Levy, H. L. and S. E. Waisbren 1987. The safety of aspartame. *Journal of the American Medical Association* 258:205.

Lewis, S. A., I. C. Lyon, and R. B. Elliott 1985. Outcome of pregnancy in the rat with mild hyperphenylalinaemia and hypertyrosinaemia: Implications for the management of "human maternal PKU." *Journal of Inherited and Metabolic Diseases* 8:113–117.

Ley, H. L. 1968. Address to the 12th Annual Food and Drug Law Institute, FDA Educational Conference, December 3, p.4.

Ley, H. L. 1969. Cited by *The New York Times* December 30.

Lieblich, I. and E. Cohen et al. 1983. Morphine tolerance in genetically selected rats induced by chronically elevated saccharin intake. *Science* 221:871–873.

Lindow, T. E. 1988. What is quality care? *New England Journal of Medicine* 318:859.

Lund, E. D., C. L. Kirkland, and P. E. Shaw 1981. Methanol, ethanol, and acetylaldehyde contents of citrus products. *Journal of Agriculture Food Chemistry* 29:361–366.

Lynch v. Merrell Dow National Laboratories, 646 F. Supp. 856, 866–67 (D. Mass. 1986), *aff'd*, 830 F. 2d 1190 (lst. Cir.1987).

MacDonald v. Ortho Pharmaceutical Corporation, 475 N.E.2d 65 (Mass. 1985).

Maher, T. J. and P. J. Kiritsy, 1987. Aspartame administration decreases the entry of a-methyldopa into the brain of rats. In *Proceedings of the First International Meeting on Dietary Phenylalanine and Brain Function*, edited by R. J. Wurtman and E. RitterWalker, Washington, D.C., May 8–10, pp.467–472.

Maher, T. J. and J. M. B. Pinto 1987. Aspartame, phenylalanine, and seizures in experimental animals. In *Proceedings of the First International Meeting on Dietary phenylalanine and Brain Function*, edited by R. J. Wurtman and E. Ritter-Walker, Washington D. C., May 8–10, pp.161–172.

Maller, O., M. R. Kare, M. Welt, and H. Bohrman 1967. Movement of glucose and sodium chloride from the oropharyngeal cavity to the brain. *Nature* 213:713.

Man, E. H., M. E. Sandhouse, J. Burg, and G. H. Fisher 1983. Accumulation of D-aspartic-acid with age in the human brain. *Science* 220:1407–1408.

Man, E. H. and J. L. Bada 1987. Dietary D-amino acids. *Annual Review of Nutrition* 7:209–225.

Martensson, E., U. Olofsson, and A. Heath 1988. Clinical and metabolic features of ethanol-methanol poisoning in chronic alcoholics. *The Lancet* 1:327–328.

Matalon, R., K. Michals, D. Sullivan, and P. Levy 1987. Aspartame consumption in normal individuals and carriers for phenylketonuria (PKU). In *Proceedings of the First International Meeting on Dietary Phenylalanine and Brain Function*, edited by R. J. Wurtman and E. Ritter-Walker, Washington D.C., May 8–10, pp.81–93.

Matsuzawa, Y. and Y. O'Hara 1984. Tissue distribution of orally administered isotopically labeled aspartame in the rat. In *Aspartame: Physiology and Biochemistry*, edited by L. D. Stegink and L. J. Filer, Jr., Marcel Dekker, Inc., New York, pp.161–199.

McEwen v. Ortho Pharmaceutical Corp., 528 P.2d 522 (Or. 1974).

McLain, D. 1986. Cited by *The Miami Herald* October 23, p.A–15.

McLean, D. R., H. Jacobs, and B. W. Mielke 1980. Methanol poisoning: a clinical and pathological study. *Annals of Neurology* 8:161.

Mehigan, D. 1981. A controlled trial of controlled trials. *New England Journal of Medicine* 305:347.

Mellanby, E. 1951. The chemical manipulation of foods. *British Medical Journal* October 13, p.864.

Menne, F. R. 1938. Acute methyl alcohol poisoning: a report of 22 incidences with post-mortem examination. *Archives of Pathology* 26:77.

Merkel, A. D., M. J. Wayner, F. B. Jolicoeur, and R. B. Mintz 1979. Effects of glucose and saccharin solutions on subsequent food consumption. *Physiological Behavior* 23:791–793.

Metzenbaum, H. 1985. Discussion of S.1557 (Aspartame Safety Act). *Congressional Record-Senate* 1985; August 1, p.S 10820.

Metzenbaum, H. M. 1988. Statement for Committee on Labor and Human Resources, *NutraSweet: Health and Safety Concerns* United States Senate, November 3, 1987. 83–178, U.S. Government Printing Office, Washington, pp.1–3.

Miller, A. B. and G. R. Howe 1977. Artificial sweeteners and bladder cancer. *Lancet* 2:1221.

Mills, J. L. 1987. Reporting provocative results: can we publish "hot" papers without getting burned? *Journal of the American Medical Association* 258:3428–3429.

Mintz, M. and J. S. Cohen 1971. *America, Inc.: Who Owns and Operates the United States*. Dell Publishing Company, New York.

Monte, W. C. 1984. Aspartame: Methyl alcohol and the public health. *Journal of Applied Nutrition* 36:42–54.

Morton, J. H. 1950. Premenstrual tension. *American Journal of Obstetrics and Gynecology* 60:343.

Morton, J. H., H. Addison, R. G. Addison, L. Hunt, and J. J. Sullivan 1953. A clinical study of premenstrual tension. *American Journal of Obstetrics and Gynecology* 65:1182.

Moser, R. H. 1969. *Diseases of Medical Progress: A Study of Iatrogenic Disease.* 3rd edition. Charles C Thomas, Springfield.

Murphy, C. 1987. Effects of age and biochemical status on preference for amino acids. In *Olfaction and Taste IX*, ed by Stephen D. Roper and Jelle Atema, *Annals of the New York Academy of Sciences* 510:515–518.

Nader, R. 1971. Introduction to *America, Inc.: Who Owns and Operates the United States*, by M. Mintz and J. S. Cohen, Dell Publishing Company, New York, pp.xi–xix.

Nehmer v. United States Veterans Administration, F. Supp. WL 52821 (N. D. Cal. 1989).

Nelson, G. 1972. Food Protection Act of 1972. *Congressional Record-Senate.* February 14, No. 18.

Neuberger, A. 1948. The metabolism of D-amino acids in mammals. *Biochemical Society Symposium* 1:20–32.

Ninomiya, Y., T. Higashi, T. Mizukoshi, and M. Funakoshi 1987. Genetics of the ability to perceive sweetness of D-phenylalanine in mice. In *Olfaction and Taste IX*, ed by Stephen D. Roper and Jelle Atema, *Annals of the New York Academy of Sciences* 510:527529.

Nisperos-Carriedo, M. D. and P. E. Shaw 1989. Comparison of volatile flavor components in fresh and processed orange juices. Presented at Annual Meeting, Institute of Food Technology, Chicago, June 25–29.

Nolley, K. and J. Nolley, 1987. Of science and ideology: a reply to Ephraim Kahn. *The Journal of Pesticide Reform* 7:34–37.

Noren, C. J., S. J. Antony-Cahill, M.. C. Griffith, and P. G. Schultz 1989. A general method for site—specific incorporation of unnatural amino acids into proteins. *Science* 244:182–188.

Novick, N. L. 1985. Aspartame-induced granulomatous panniculitis. *Annals of Internal Medicine* 102:206–207.

Olney, J. W. and 0. L. Ho 1970. Brain damage in infant mice following oral intake of glutamate, aspartame or cysteine. *Nature* 227: 609–611.

Olney, J. W. 1979. In *Glutamic Acid: Advances in Biochemistry and Physiology*, edited by L.J. Filer, Raven, New York, p.287.

Olney, J. W., J. Labruyere, and T. de Gubareff 1980. Brain damage in mice from voluntary ingestion of glutamate and aspartame. *Neurobehavioral Toxicology* 2:125–129.

Pardridge, W. M. 1986. The safety of aspartame. *Journal of the American Medical Association*. 256:2678.

a. Pardridge, W. M. 1987. The safety of aspartame. *Journal of the American Medical Association* 258:206.

b. Pardridge, W. M. 1987. Phenylalanine transport at the human blood-brain barrier. In *Proceedings of the First International Meeting on Dietary Phenylalanine and Brain Function*, edited by R. J. Wurtman and E. Ritter-Walker, Washington, D.C., May 8–10, pp.57–69.

Pasteur, L. 1852. Untersuchungen uber Asparaginsaurer und Aepfelsaure. *Annals Chemistry* (German). 82:324–335.

Pearce, J. M. S. 1988. Exploding head syndrome, *The Lancet* 2:270–271.

Pellegrino, E. 1987. Cited by *The Miami Herald* February 5, p.B–1.

Petsko, G. A. and D. Ringe 1988. Blueprints for protein engineering. Presented at the annual meeting of the American Association for the Advancement of Science, Boston, February 13.

Phillip v. Kimwood Machine Co., 269 Or. 485, 525, P.2d 1033 (1974).

Pickford, J. C., E.. H. McGale, and G. M. Aber 1973. Studies on the metabolism of phenylalanine and tyrosine in patients with renal disease. *Clinica Chimica Acta* 48:77–83.

Pincus, J. and K. Barry 1986. Feature in *Medical Tribune* August 27, p.19.

Pincus, J. H., et al. 1987. Plasma levels of amino acids correlate with motor fluctuations in Parkinsonism. *Archives of Neurology* 44: 1006–1009.

Pi-Sunyer, F. X. 1988. Testimony for the Committee on Labor and Human Resources, U.S. Senate, Hearing on *"NutraSweet"—Health and Safety Concerns* November 3, 1987. 83–178, U.S. Government Printing Office, Washington, pp.390–392.

Posner, H. S. 1975. Biohazards of methanol in proposed new uses. *Journal of Toxicology and Environmental Health* 1:153–171.

Price, J. M., C. G. Biava, and B. L. Oser, et al. 1970. Bladder tumors in rats fed cyclohexylamine or high doses of a mixture of cyclamate and saccharin. *Science* 167:1131.

Procter, A. W., A. M. Palmer, G. C. Stratmann, and D. M. Bowen, 1986. Glutamate/aspartame-releasing neurones in Alzheimer's disease. *New England Journal of Medicine* 314:1711–1712.

Randolph, T. G. 1956. The descriptive features of food addiction: addictive eating and drinking. *Quarterly Journal of Studies on Alcohol* 17:198–224.

Rao, K. R., A. L. Aurora, S. Mithaiyan, and S. Ramakrishnan 1977. Biochemical changes in brain in methanol poisoning—an experimental study. *Indian Journal Of Medical Research* 65:285–292.

Reitano, G., G. Distefano, and R. Vigo, et al. 1978. Effect of priming of amino acids on insulin and growth hormone response in the premature infant. *Diabetes* 27:334–337.

Rennie, D., E. Knoll, and A. Flanagin 1989. The international congress on peer review in biomedical publication. *Journal of the American Medical Association* 261:749.

Ribicoff, A. S. 1971. *Chemicals and the Future of Man.* The U.S. Senate Subcommittee on Executive Reorganization and Government Research, Committee on Government Operations, April 6, pp.1–2.

Richardson v. Richardson-Merrell, Inc., 857 F.2d 823 (D.C. Cir. 1988, *Petition for rehearing and rehearing en banc* October 26, 1988).

Roberts, H. J. 1960. Long-term effective weight reduction: Experiences with Metrecal. *American Journal of Clinical Nutrition* 8:817.

Roberts, H. J. 1962. Long-term weight reduction in cardiovascular disease. Experiences with a hypocaloric food mixture (Metrecal) in 78 patients. *Journal of the American Geriatrics Society* 10:308.

a. Roberts, H. J. 1964. The syndrome of narcolepsy and diabetogenic hyperinsulinism in the American Negro: Its relationship to diabetes mellitus, obesity, dysrhythmias and accelerated cardiovascular disease. *Journal of the National Medical Association* 56:18.

b. Roberts, H. J. 1964. The syndrome of narcolepsy and diabetogenic ("functional") hyperinsulinism, with special reference to obesity, diabetes, idiopathic edema, cerebral dysrhythmias and multiple sclerosis (200 patients). *Journal of the American Geriatrics Society* 12:926–976.

c. Roberts, H. J. 1964. Chronic refractory fatigue—an "organic" perspective: With emphasis upon the syndrome of narcolepsy and diabetogenic hyperinsulinism. *Medical Times* 92:1144–1160.

d. Roberts, H. J. 1964. Afternoon glucose tolerance testing: A key to the pathogenesis, early diagnosis and prognosis of diabetogenic hyperinsulinism. *Journal of the American Geriatrics Society* 12:423–472.

e. Roberts, H. J. 1964. Fatigue as an elusive organic problem. *Consultant* 4:30 (May).

a. Roberts, H. J. 1965. Spontaneous leg cramps and "restless legs" due to diabetogenic hyperinsulinism: observations on 131 patients. *Journal of the American Geriatrics Society* 13:602–638.

b. Roberts, H. J. 1965. Diabetogenic hyperinsulinism—a major etiology of ischemic heart disease. *Clinical Research* 13:28.

c. Roberts, H. J. 1965. Café-au-lait spots (CALS), localized hypomelanosis (LH), and the white forelock (WF)—clues to the syndrome of narcolepsy and diabetogenic hyperinsulinism. *Clinical Research* 13:267.

d. Roberts, H. J. 1965. Triparanol myxedema. *Journal of the American Geriatrics Society* 13:520–256.

a. Roberts, H. J. 1966. The role of diabetogenic hyperinsulinism in the pathogenesis of prostatic hyperplasia and malignancy. *Journal of the American Geriatrics Society* 14:795.

b. Roberts, H. J. 1966. An inquiry into the pathogenesis, rational treatment and prevention of multiple sclerosis, with emphasis upon the combined role of diabetogenic hyperinsulinism and recurrent edema. *Journal of the American Geriatrics Society* 14:586–608.

c. Roberts, H. J. 1966. On the etiology, rational treatment and prevention of multiple sclerosis. *Southern Medical Journal* 59:940.

a. Roberts, H. J. 1967. The role of diabetogenic hyperinsulinism in nocturnal angina pectoris, with special reference to the etiology of ischemic heart disease. *Journal of the American Geriatrics Society* 15:545–555.

b. Roberts, H. J. 1967. Obesity due to the syndrome of narcolepsy and diabetogenic hyperinsulinism: Clinical and therapeutic observations on 252 patients. *Journal of the American Geriatrics Society* 15:721.

c. Roberts, H. J. 1967. Migraine and related vascular headaches due to diabetogenic hyperinsulinism: Observations on pathogenesis and rational treatment in 421 patients. *Headache* 7:41–62.

a. Roberts, H. J. 1968. The value of afternoon glucose tolerance testing in the diagnosis, prognosis and rational treatment of "early chemical diabetes": A 5-year experience. *Acta Diabetologica Latina* 5:532.

b. Roberts, H. J. 1968. Are the massive diet-fat-heart and coronary drug studies justified? A critical commentary. *Angiology* 19:652.

a. Roberts, H. J. 1969. Hyperthyroidism and thyroiditis precipitated by severe caloric restriction: a report of 8 cases. Abstract 305. Program of the 51st meeting, Endocrine Society, New York, June 27.

b. Roberts, H. J. 1969. A clinical and metabolic re-evaluation of reading disability. *In Selected Papers on Learning Disabilities. 5th annual conference, Association for Children with Learning Disabilities.* Academic Therapy Publications, San Raphael, California, p.472–490.

c. Roberts, H. J. 1969. Oral therapy in "early chemical diabetes." I.Serial glucose, cholesterol and uric acid responses to phenformin. *Acta Diabetologica Latina* 6:728.

Roberts, H. J. 1970. Unrecognized narcolepsy and amphetamine abuse. *Medical Counterpoint* 2:28.

a. Roberts, H. J. 1971. The role of pathologic drowsiness in traffic accidents: an epidemiologic study. *Tufts Medical Alumni Bulletin* 32:21–30.

b. Roberts, H. J. 1971. *The Causes, Ecology and Prevention of Traffic Accidents.* Charles C Thomas, Springfield.

Roberts, H. J. 1973. Spontaneous leg cramps and "restless legs" due to diabetogenic (functional) hyperinsulinism. *Journal of the Florida Medical Association* 60(5):29–31.

Roberts, H. J. 1979. *Is Vasectomy Safe? Medical, Public Health and Legal Implications.* Sunshine Academic Press.

Roberts, H. J. 1985. The hazards of very-low-calorie dieting. *American Journal of Clinical Nutrition* 41:171–172.

Roberts, H. J. 1986. Respiratory hazards of anti-static clothes softeners. *Palm Beach County Medical Society Bulletin* January, pp.24–31.

a. Roberts, H. J. 1987. Is aspartame (NutraSweet®)safe? *On Call* (Palm Beach County Medical Society) January, pp.16–20.

b. Roberts, H. J. 1987. Neurologic, psychiatric and behavioral reactions to aspartame in 505 aspartame reactors. In *Proceedings of the First International Conference on Dietary Phenylalanine and Brain Function*, edited by R.J. Wurtman and E. Ritter-Walker, Washington, D.C., May 8–10, pp.477–481.

a. Roberts, H. J. 1988. Aspartame (NutraSweet®)-associated confusion and memory loss: A possible human model for early Alzheimer's disease. Abstract 306. Annual meeting of the American Association for the Advancement of Science, in Boston, February 13.

b. Roberts, H. J. 1988. Aspartame (NutraSweet®)-associated epilepsy. *Clinical Research* 36:349A.

c. Roberts, H. J. 1988. Complications associated with aspartame (NutraSweet®) in diabetics. *Clinical Research* 3:489A.

d. Roberts, H. J. 1988. The Aspartame Problem. Statement for Committee on Labor and Human Resources, U.S. Senate, Hearing on *"NutraSweet"— Health and Safety Concerns* November 3, 1987. 83–178, U.S. Government Printing Office, Washington, pp.466–467.

e. Roberts, H. J. 1988. Reactions attributed to aspartame-containing products: 551 cases. *Journal of Applied Nutrition* 40:85–94.

a. Roberts, H. J. 1989. Public comment submitted to FDA regarding GRASP 8G0345—microparticulated egg and milk protein product. February 17.

b. Roberts, H. J. 1989. New perspectives concerning Alzheimer's Disease. *On Call.* (Palm Beach County Medical Society) August, pp.14–16.

Roberts, H. J.: *Reactions to Aspartame: A Clinical, Pharmacologic and Public Health Inquiry.* To be submitted for publication.

Roe, 0. 1982. Species differences in methanol poisoning. *CRC Critical Reviews in Toxicology* October, pp.275–286.

Rogers, T. R. and P. B. M. Leung 1973. The influence of amino acids on the neuroregulation of food intake. *Federation Proceedings* 32:1709–1719.

Rossi, A. C, L. Bosco, and G. A. Faich, et al. 1988. The importance of adverse reaction reporting by physicians. *Journal of the American Medical Association* 259:1203–1204.

Sapira, J. D. 1988. Which will be the best medical school in ten years? *Southern Medical Journal* 81:1079.

Schenebeck v. Sterling Drug, Inc., 423 F.2d 919 (1970).

Schiedermayer, D. L. and M. Siegler 1986. Believing what you read: Responsibilities of medical authors and editors. *Archives of Internal Medicine* 146:2043–2044.

Schmid, R., V. Schuszdziarra, E. Schulte-Frohlinde, V. Marer, and M. Classen 1989. Circulating amino acids and pancreatic endocrine function after ingestion of a protein-rich meal in obese subjects. *Journal of Endocrinology and Metabolism* 68:1106–1110.

Schrage, M. 1988. Are ideas viruses of the mind? *The Miami Herald* November 13, pp.C-l, C-6.

Schultz, W. B. 1989. Cancer-causing substances in food, drugs and cosmetics. (Letter) *New England Journal of Medicine* 320:934–935.

Sczuc, E. F., K. E. Barrett, and D. D. Metcalfe 1986. The effects of aspartame on mast cells and basophils. *Food and Chemical Toxicology* 24:171–174.

Shabin, H. M. and M. L. Albert 1988. Aspartame: An evaluation of adverse effects. *Hospital Formulary* 23:543–546.

Shah, J., H. Carlson, M. Peters, and J. Carr 1986. Differences in amino acid-induced insulin release between male and female. Abstract 308. 68th' annual meeting of the Endocrine Society, Anaheim, California, June 24–27.

Shipp, J. C., A. E. Matos, H. Knizley, and L. E. Crevasse 1964. C02 formed from endogenous and exogenous substrates in perfused rat heart. *American Journal of Physiology* 207:1231.

Sindell v. Abbott Laboratories, 26 Ca1.3d 588, 607 P.2d 924, 163 Cal.Rpt. 132, cert. denied, 449 U. S. 912, 101 S.Ct 285, 66 L.Ed.2d 140 (1980).

Sokoloff, L. 1967. Action of thyroid hormones and cerebral development. *American Journal of Diseases of Children* 114:498–506.

Solzhenitsyn, A. I. 1969. Letter, *The New York Times* November 14.

Sommer, H. 1962. The physiological fate of methyl alcohol released from pectin. *Industrielle Obst. und Gemeseverwesting* 47:172–173.

Spellacy, W. N. and K. L. Carlson 1966. Plasma insulin and blood glucose levels in patients taking oral contraceptives. *American Journal of Obstetrics and Gynecology* 95:474.

Spencer, P.S., et al. 1987. Guam amyotrophic lateral sclerosis parkinsonism-dementia linked to a plant excitant neurotoxin. *Science* 237:517–522.

Spruill v. Boyle-Midway, Inc., 308 F.2d 79, 85 (4 Cir. 1962).

Staamp, J. 1929. *Some Economic Factors in Modern Life*. P. S. King & Son, Ltd., London, pp.258–259.

Staff, J., D. Jacobson, C. R. Tillman, C. Curington, and P. Toskes 1984. Protease-specific suppression of pancreatic exocrine secretion. *Gastroenterology* 87:44–52.

Steqink, L. D. and L. J. Filer, Jr. 1984. *Aspartame: Physiology and Biochemistry*, Marcel Dekker, Inc., New York.

Steqink, L. D., L. J. Filer, Jr., E. F. Bell, and E. E. Ziegler 1987. Plasma amino acid concentrations in normal adults administered aspartame in capsules or solution: Lack of bioequivalence. *Metabolism* 36:507–512.

St. George-Hyslop, P.H., R. E. Tanzi, and R. J. Polinsky, et al. 1987. Absence of duplication of chromosome 21 genes in familial and sporadic Alzheimer's disease. *Science* 238:664–666.

Stonier, C., E. H. McGale, and G. M. Aber 1984. Studies of phenylalanine hydroxylase activity in patients with chronic renal failure: the effect of hemodialysis. *Clinica Chimica Acta* 143:115– 122.

Sturtevant, F. M. 1985. Aspartame—a new ingredient: Reply to the critical comments of Woodrow C. Monte. *Journal of Environmental Science and Health* 20:863–901.

Swartz, R. D., R. P. Millman, and J. E. Billi, et al. 1981. Epidemic methanol poisoning: Clinical and biochemical analysis of a recent episode. *Medicine* 60:373–382.

Tampa Drug Co. v. Waite, 103 So. 2d 609 (Fla. 1958).

Taylor, L. 1988. Testimony for Committee on Labor and Human Resources, U.S. Senate, Hearing on *"NutraSweet"—Health and Safety Concerns* November 3, 1987. 83–178, U.S. Government Printing Office, Washington, pp.303–305.

Tephly, T. R. and K. E. McMartin 1984. Methanol metabolism and toxicity. In *Aspartame: Physiology and Biochemistry*, edited by L. D. Steqink, J. L. Filer, Jr., Marcell Dekker, Inc., New York, pp.111–140.

Tocci, P. M. and B. Beber 1973. Anomalous phenylalanine loading responses in relation to cleft lip and cleft palate. *Pediatrics* 52:109.

Tollefson, L., R. J. Barnard, and W. H. Glinsmann 1987. Monitoring of adverse reactions to aspartame reported to the U.S. Food and Drug Administration. In *Proceedings of the First International Meeting on Dietary Phenylalanine and Brain Function*, edited by R. J. Wurtman and E. Ritter-Walker, Washington, D.C., May 8–10, pp.347–372.

Tourian, A. 1985. Control of phenylalanine hydroxylase synthesis and tissue culture by serum and insulin. *Journal of Cellular Physiology* 87:15–24.

Turski, L., D. S. Meldrum, and E. A. Cavalheiro et al. 1984. Paradoxical anticonvulsant activity of the excitatory amino acid N-methyl-D-aspartate in the rat caudate-putamen. *Proceedings of the National Academy of Sciences USA* 64:1689–1693.

Uribe, M. 1982. Potential toxicity of a new sugar substitute in patients with liver disease. *New England Journal of Medicine* 306:173.

Verrett, J. 1988. Testimony for the Committee on Labor and Human Resources, U.S. Senate, Hearing on *"NutraSweet"—Health and Safety Concerns* November 3, 1987. 83–178, U.S. Government Printing Office, Washington, p.385.

Vianna, N. J. 1975. Hodgkin's disease. *Journal of the American Medical Association* 234:1133.

Vinters, H. V., B. L. Miller, and W. M. Pardridge 1988. Brain amyloid and Alzheimer's disease. *Annals of Internal Medicine* 109:41–54.

Virkkunen, M. 1982. Reactive hypoglycemic tendency amoung habitually violent offenders: A further study by means of the glucose tolerance test. *Neuropsychobiology* 8:35–40.

Virkkunen, M. 1983. Insulin secretion during the glucose tolerance test in antisocial personality. *British Journal of Psychiatry* 142:598–604.

Virkkunen, M. 1984. Reactive hypoglycemic tendency amoung arsonists. *Divica Scandinavica* 69:445–452.

Virkkunen, M., A. Nuutioa, F. K. Goodwin, and M. Linnoila, M. 1987. CSF monoamine metabolities in arsonists. *Archives of General of Psychiatry* 44;241–247.

Wallace, D. and F. Wallace 1988. Cited in *People* Magazine November 28, p.159.

Walton, R. G. 1986. Seizure and mania after "high intake of aspartame. *Psychosomatics* 27:218–220.

Wang, H. L. and H. A. Waisman 1961. Phenylalanine tolerance tests in patients with leukemia. *Journal of Laboratory and Clinical Medicine* 57:73–77.

Wannemacher, R. W., A. S. Klainer, R. E. Dinterman, and W. B. Beisel 1976. The significance and mechanism of an increased serum phenylalaninetyrosine ratio during infection. *American Journal of Clinical Nutrition* 29:997–1006.

Way, E. L. and R. Hausman 1950. Effect of tetra-ethyl thiuramdisulfide (Antabuse) on toxicity of methyl alcohol. *Federation Proceedings* 9:324.

Weber, J. C. P. and J. P. Griffin 1986. Adverse reactions in the elderly. *The Lancet* 2:291.

Weiffenbach, J. M., P. C. Fox, and B. J. Baum 1987. Taste and salivary gland dysfunction. In *Olfaction and Taste IX*, ed by Stephen D. Roper and Jelle Atema, *Annals of the New York Academy of Science* 510:698–699.

Weiss, J. H. and D. W. Choi 1988. Beta-N-methylamino-L-alanine neurotoxicity: Requirement for bicarbonate as a cofactor. *Science* 241:973–975.

Weissler, A. M. and F. A. Kruger 1964. Effect of glucose on the performance of the hypoxic isolated rat heart. (Abstract) *Circulation* 29 (III:117.

Weidner, G. and J. Istvan 1985. Dietary sources of caffeine. *New England Journal of Medicine* 13:1421.

Wells v. Ortho Pharmaceutical Corp., 788 F.2d 741, *reh'g denied*, 795 F.2d (llth Cir.), *Cert. denied*, 479 U.S. 950 (1986).

Wernicke, J. F. 1985. The side effect profile and safety of fluoxetine. *Journal of Clinical Psychiatry* 46:59–67, 1985.

Wisconsin Alumni Research Foundation 1973. *Long Term Saccharin Feeding in Rats. Final Report.* Madison, WARF.

Wurtman, R. J. 1985. Aspartame. Possible effect on seizure susceptibility. *The Lancet* 2:1060.

Wurtman, R. J. 1986. Press conference on Cable News Network (CNN), July 17.

a. Wurtman, R. J. 1987. Aspartame effects on brain serotonin. *American Journal of Clinical Nutrition* 45:799–801.

b. Wurtman, R. J. 1987. Cited by *The Philadelphia Inquirer* February 22, p.H–9.

Yuhas v. Mudge, 129 N.J. Super. 207, 322 A2d 824 (1974).

Zeytinoglu, I. Y., C. N. Gherondache, and G. Pincus 1969. The process of aging: Serum glucose and immunoreactive insulin levels during the oral glucose tolerance test. *Journal of the American Geriatrics Society* 17:1, 1969.

INDEX